全国机械行业职业教育优质规划教材（高职高专）
经全国机械职业教育教学指导委员会审定

机械制造专业英语

第2版

ENGLISH COURSE FOR MECHANICAL ENGINEERING

主　编　王晓江
参　编　吴　兵　魏康民　卢端敏
张兆隆　王靖东　李学哲
主　审　田锋社

机械工业出版社

本书是全国机械行业职业教育优质规划教材，经全国机械职业教育教学指导委员会审定。

全书分为10章，共38个单元。内容包括机械加工基础知识，工程材料及热处理，工程材料成形方法，刀具、夹具和量具，机床，机械加工工艺，机械CAD/CAM与数控机床，电加工技术，先进制造技术，加工质量检测与控制技术等方面。每个单元由课文、问题、单词和词组、难句分析、专业术语（800余条）和阅读材料等部分组成。本书课文和阅读材料均节选自英文原版教材和相关专业英文资料，注重反映专业领域的新技术、新工艺和新材料，内容全面、图文并茂、难易适中，融知识性和趣味性于一体，使读者在掌握机械制造专业英语的同时进一步拓展机械制造专业的相关知识。为了便于学习，本书在附录中还依次介绍了相关专业标准的中英文对照、专业英语的翻译技巧、常用计量单位的表示方法、英文中数字和日期的表示形式等内容。

本书为高职高专机械制造与自动化专业学生的专业英语教材，也可作为机械设计制造类其他专业及相近专业的教学参考书，同时还可供机械制造行业企业的工程技术人员学习和参考。

本书配有电子课件，凡使用本书作为教材的教师均可登录机械工业出版社教育服务网（http：//www.cmpedu.com）下载，或发送电子邮件至cmpgaozhi@sina.com索取。咨询电话：010-88379375。

图书在版编目（CIP）数据

机械制造专业英语/王晓江主编．—2版．—北京：机械工业出版社，2017.9（2023.8重印）
全国机械行业职业教育优质规划教材（高职高专）
经全国机械职业教育教学指导委员会审定
ISBN 978-7-111-56964-0

Ⅰ.①机… Ⅱ.①王… Ⅲ.①机械制造—英语—高等职业教育—教材 Ⅳ.①TH

中国版本图书馆CIP数据核字（2017）第120015号

机械工业出版社（北京市百万庄大街22号　邮政编码100037）
策划编辑：刘良超　责任编辑：王　丹　刘良超
责任校对：刘　岚　封面设计：陈　沛
责任印制：任维东
北京圣夫亚美印刷有限公司印刷
2023年8月第2版第7次印刷
184mm×260mm · 16.5印张 · 401千字
标准书号：ISBN 978-7-111-56964-0
定价：48.00元

电话服务
客服电话：010-88361066
010-88379833
010-68326294

网络服务
机　工　官　网：www.cmpbook.com
机　工　官　博：weibo.com/cmp1952
金　书　网：www.golden-book.com
机工教育服务网：www.cmpedu.com

第2版前言

本书是全国机械行业职业教育优质规划教材，是根据全国机械职业教育教学指导委员会确定的全国机械行业职业教育优质规划教材立项要求和“机械制造专业英语”课程教学标准编写的。

本书第1版自2009年出版以来，先后重印8次，印数超过20000册，深受广大高职高专院校师生的欢迎和好评。此次修订基于第1版内容，根据机械设计制造类专业的发展和行业企业对高素质技术技能人才专业英语水平的要求，征求了部分使用本书第1版的院校教师的意见，听取了部分制造类企业工程技术人员对本书的建议，主要进行了以下几个方面的修订：

1）为使本书的专业针对性更强，对原书部分章节进行了适当的调整和删减，对课文内容、配图做了适当的补充和完善。

2）为了拓展专业知识和方便学习，附录部分增加了 Tables of Weights and Measures 和 Numerals，Days and Months 两部分内容。

3）对原书中部分单元的专用词组进行了适当的增减。

本书修订后仍为10章，共38个单元。各校在组织教学时，可根据专业和学生的实际情况进行适当的选取和调整。

本书由陕西工业职业技术学院王晓江主编，参加编写的人员还有陕西工业职业技术学院吴兵、魏康民，张家界航空职业技术学院卢端敏，河北机电职业技术学院张兆隆，沈阳职业技术学院李学哲和包头职业技术学院王靖东。陕西工业职业技术学院田锋社审阅了本书，并提出了宝贵意见。

本书的修订还得到了陕西工业职业技术学院杜和平，河北机电职业技术学院张敬芳，包头职业技术学院邢志刚、张桂霞等老师的大力支持，在此谨向他们表示衷心的感谢！

由于编者水平有限，书中难免会有缺点和错误，敬请广大读者批评指正。

编　者

第1版前言

本书是全国机械职业教育机械制造与自动化专业教学指导委员会规划教材、高等职业教育机电类规划教材。本书是根据“机械制造与自动化专业英语”课程教学大纲编写的，其目的是为了更好地帮助机械制造专业学生进一步适应本专业国际、国内发展的需要，提高直接阅读英语原文和翻译有关专业英语书刊的能力，学习和借鉴国外先进的制造技术，从而推进我国机械制造行业的发展。本书内容大部分节选自英、美等国专业教材及专业刊物。全书共分10章，38个单元。内容涉及机械加工基础知识，工程材料及热处理，工程材料成形方法，刀具、夹具和量具，机床，机械加工工艺，CAD/CAM与数控机床，电加工技术，先进制造技术，质量检测与控制技术等。

本书可供高职高专机械制造与自动化专业学生使用，也可供有关机械制造类企业的工程技术人员参考。在实际教学中，各院校可根据实际情况调整授课顺序或删减有关内容。

本书由陕西工业职业技术学院王晓江主编（编写第1、2、3章），参加编写的人员还有陕西工业职业技术学院吴兵（编写第4、5章）、魏康民（编写第10章），张家界航空职业技术学院卢端敏（编写第8、9章），河北机电职业技术学院张兆隆（编写第6章），沈阳职业技术学院李学哲（编写附录A、B），包头职业技术学院王靖东（编写第7章）。本书由陕西工业职业技术学院田锋社教授主审，陕西工业职业技术学院澳大利亚籍教师Paul Conroy审阅了全书。

本书在编写过程中得到了张普礼、殷城、侯会喜、钱泉森等同志的大力支持，徐惠、肖春艳老师对教材提出了许多宝贵的修改意见，在此一并表示衷心的感谢。

由于编者水平有限，加上时间紧迫、经验不足，书中难免会有缺点和错误，欢迎读者批评指正。

编　者

CONTENTS

Chapter 1　Fundamentals of Machine Manufacturing

Unit 1　Third-Angle Projection

Text

The six *views*. Any object can be viewed from six *mutually perpendicular* directions, as shown in Figure 1-1-1a. These six views may be *drawn* if necessary, and they are always *arranged* as shown in Figure 1-1-1b, which is the American National Standard arrangement. The *top*, *front*, and bottom views *align* vertically, while the *rear*, left-side, front, and right-side views align horizontally. To draw a view out of place is a serious error and is generally regarded as one of the worst possible mistakes in drawing①.

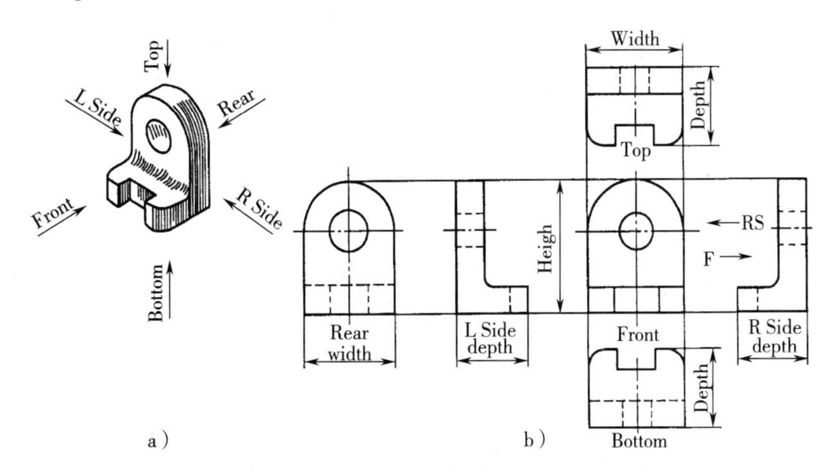

Figure 1-1-1　The six views

Note that height is shown in the rear, left-side, front, and right-side views; width is shown in the rear, top, front, and bottom views; and depth is shown in the four views that surround the front view—namely, the left-side, top, right-side, and bottom views. Each view shows two of the principal dimensions. Note also that in the four views that surround the front view, the front of the object faces toward the front view.

Adjacent views are *reciprocal*. If the front view in Figure 1-1-1 is imagined to be the object itself, the right-side view is obtained by looking toward the right side of the front view, as shown by the *arrow* RS. Likewise, if the right-side view is imagined to be the object, the front view is obtained by looking toward the left side of the right-side view, as shown by the arrow F. The same relation exists between any two adjacent views.

Necessary views. A drawing for use in production should contain only those views needed for a clear and complete shape *description* of the object. These minimum required views are referred to as the necessary views. In selecting views, the drafter should choose those that best show essential contours or shapes and have the least number of hidden lines.

As shown in Figure 1-1-1, three *distinctive* features of this object need to be shown on the drawing: (1) rounded top and hole, seen from the front; (2) rectangular notch and rounded corners, seen from the top; (3) right angle with filleted corner, seen from the side.

The three principal dimensions of an object are width, height, and depth. In technical drawing, these fixed terms are used for dimensions taken in these directions, regardless of the shape of the object②. The terms "length" and "thickness" are not used because they cannot be applied in all cases. The top, front, and right-side views, arranged close together, are shown in Figure 1-1-1. These are called the three regular views because they are the views most frequently used.

Alignment of views. Errors in arranging the views are so commonly made by students that it is necessary to repeat this: the views must be drawn in accordance with the American National Standard arrangement shown in Figure 1-1-1. Figure 1-1-2a shows an offset guide that requires three views. These three views, correctly arranged, are shown in Figure 1-1-2b. The right-side view must be directly to the right of the front view—not out of alignment, as shown in Figure 1-1-2c. Also, never draw the views in reversed positions, with the bottom over the front or the right-side to the left of the front view (Figure 1-1-2d), even through the views do line up with the front view.

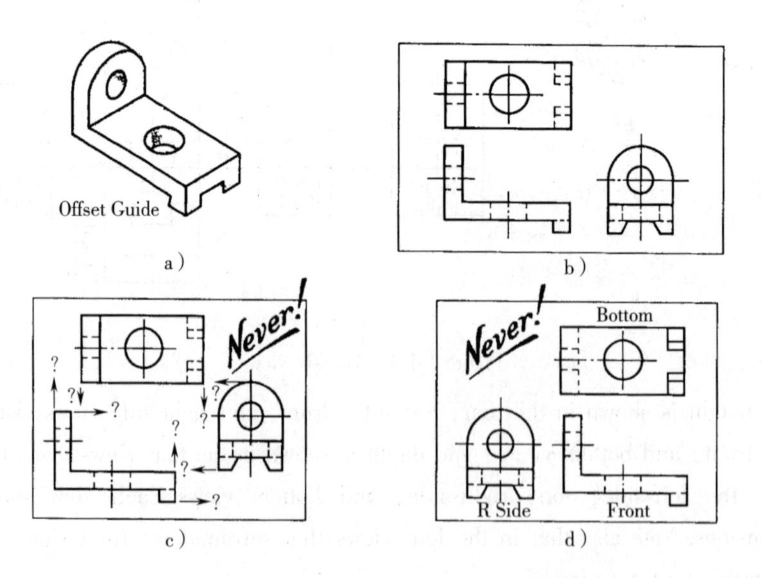

Figure 1-1-2 Position of views

Questions

1. What is the third-angle projection?

2. What are the differences between third-angle projection and first-angle projection?

3. List the six principal views of an object.

4. In a drawing that shows the top, front, and right-side view, which two views show depth? Which view shows depth vertically? Which view shows depth horizontally?

5. What are the three principal dimensions of an object?

New Words and Expressions

1. projection [prəˈdʒekʃən] n. 投影，发射
2. view [vjuː] n. 视图 vt. 观察
3. mutual [ˈmjuːtʃuəl] adj. 相互的，共同的
4. perpendicular [pəːpənˈdikjulə] adj. 垂直的，正交的 n. 垂直
5. draw [drɔː] vt. 拉，拖，绘制，描写 vi. 制图
6. arrange [əˈreindʒ] vt. 整理，排列 vi. 安排，准备
7. top [tɔp] n. 顶端，上部 adj. 最高的，顶上的
8. front [frʌnt] n. 正面，前面 adj. 正面的 vt. & vi. 面向
9. align [əˈlain] vt. & vi. 使成一直线，排列成一行
10. rear [riə] n. 后部，后面 adj. 后面的，后部的
11. adjacent [əˈdʒeisənt] adj. 接近的，毗邻的
12. reciprocal [riˈsiprəkəl] adj. 相互的，相应的 n. 倒数
13. arrow [ˈærəu] n. 箭，指针，箭号
14. description [diˈskripʃən] n. 叙述，图说，绘制
15. distinctive [diˈstiŋktiv] adj. 有区别的，特殊的
16. arrowhead [ˈærəuhed] n. 箭头
17. left-side view 左侧（视）图
18. right-side view 右侧（视）图
19. hidden line (dotted line, dashed line) 隐藏线，虚线
20. be out of (the) perpendicular 倾斜
21. from top to tail (toe) 从头到尾，整个
22. front and rear 在前后；前部和后部
23. in accordance with 按照，依据，与……一致
24. line up with 排成一行
25. be adjacent 靠近，与……邻接
26. be generally regarded as 一般地被看作……，一般地被认为……
27. look toward 面朝，期待；为了……作好准备
28. be referred to as 称为，被认为是

Notes

[1] To draw a view out of place is a serious error and is generally regarded as one of the worst possible mistakes in drawing.

将视图绘制在不适当的位置是一个严重的错误，而且通常被认为是绘图过程中可能出现的最为严重的错误之一。

句中 to draw a view out of place 为不定式短语，在句中做主语；句中 out of place 可译为“不合适，不在适当的位置”；句中 is generally regarded as 可译为“通常被认为……”。

[2] In technical drawing, these fixed terms are used for dimensions taken in these directions, regardless of the shape of the object.

在技术（专业）绘图中，不论物体的形状如何，上述固定术语都被用来表示物体在这些方向上对应的尺寸。

in technical drawing 在句中做状语，可译为“在技术（专业）制图中”；taken in these directions 是过去分词短语做后置定语，意为“在这些方向上测得的”；regardless of 作“不管；不顾；不论……如何”解。

Glossary of Terms

1. third-angle projection 第三角投影
2. first-angle projection 第一角投影
3. mechanical drawing 机械制图
4. standard drawing 标准图
5. standard components (parts) 标准件
6. drawing sheet, drawing paper 图纸
7. drawer, draftsman, drafter 绘图员
8. working drawing 工作图，生产图
9. detail drawing, part drawing 零件图
10. sketch (layout, outline) 草图
11. assembly drawing 装配图
12. design drawing 设计图
13. blueprint 蓝图
14. engineering drawing 工程图
15. structure drawing 结构图
16. machine components (parts) 零（部）件
17. title blocks 标题栏
18. sectional view 剖视图
19. orthographic projection 正投影
20. the top view 俯视投影，俯视图
21. the front view 主视投影（主视图）

22. the side view 侧投影，侧视图
23. the bottom view 仰视图
24. rear (back) view 后视图
25. end view 端视图
26. three-view drawing 三视图
27. pictorial drawing 立体图
28. profile, section (full ~, half ~, offset ~, broken-out ~, rotating ~, inclined ~, compound ~) 剖面（全剖、半剖、阶梯剖、局部剖、旋转剖、斜剖、复合剖）
29. technical requirements 技术要求
30. a detail list of components 零件明细栏
31. scale, proportional scale 比例
32. dimensional line 尺寸线
33. descriptive geometry 画法几何
34. dimensioning, size marking 标注尺寸
35. straight line (arc, curve) 直线（圆弧，曲线）
36. horizontal line (incline line, vertical line) 水平线（斜线，垂直线）
37. continuous thick line (full line, visible line) 粗实线
38. continuous thin line 细实线

Reading Materials

First-Angle Projection

If the vertical and horizontal planes of projection are considered indefinite in extent and intersecting at 90° with each other, the four dihedral angles produced are the first, second, third, and fourth angles (Figure 1-1-3a).

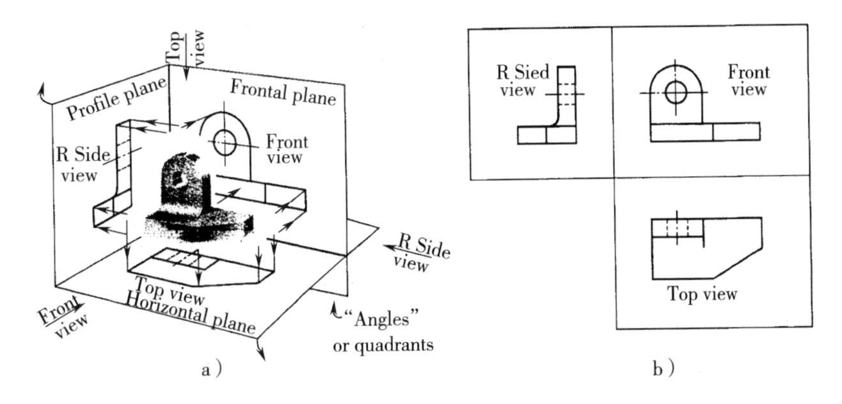

Figure 1-1-3 First-angle projection

If the object is placed above the horizontal plane and in front of the vertical plane, the object is in the first angle. In this case, the observer always looks through the object and to the planes of projection. Thus, the right-side view is still obtained by looking toward the right side of the object, the front by looking toward the front, and the top by looking down toward the top; but the views are projected from the object onto a plane in each case. When the planes are unfolded (Figure 1-1-3b), the right-side view falls at the left of the front view, and the top view falls below the front view, as shown. A comparison between first-angle orthographic projection and third-angle orthographic projection is shown in Figure 1-1-4. The front, top, and right-side views shown in Figure 1-1-3b for first-angle projection are repeated in Figure 1-1-4a. Ultimately, the only difference between third-angle and first-angle projection is the arrangement of the views. Still, confusion and possibly manufacturing errors may result when the user reading a first-angle drawing thinks it is a third-angle drawing, or vice versa. To avoid misunderstanding, international projection symbols, shown in Figure 1-1-4, have been developed to distinguish between first-angle and third-angle projections on drawings. On drawings where the possibility of confusion is anticipated, these symbols may appear in or near the title box.

In the United States and Canada (and, to some extent, in England), third-angle projection is standard, while in most of the rest of the world, first-angle projection is used. First-angle projection was originally used all over the world, including the United States, but it was abandoned around 1890.

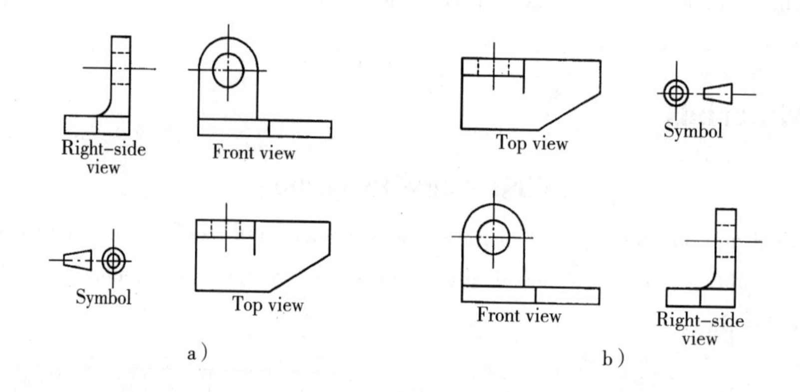

Figure 1-1-4 First-angle projection compared to third-angle projection

Lines in Sectioning

A correct front view and sectional view are shown in Figure 1-1-5a and b. In general, all visible edges and contours behind the cutting plane should be shown; otherwise a section will appear to be made up of disconnected and unrelated parts, as shown in Figure 1-1-5c. Occasionally, however, visible lines behind the cutting plane are not necessary for clarity and should be omitted.

Sections are used primarily to replace hidden-line representation; and, as a rule, hidden lines should be omitted in sectional views. As shown in Figure 1-1-5d, the hidden lines do not clarify the drawing; they tend to confuse, and they take unnecessary time to draw. Sometimes hidden lines are necessary for clarity and should be used in such cases, especially if their use will make it possible to omit a view.

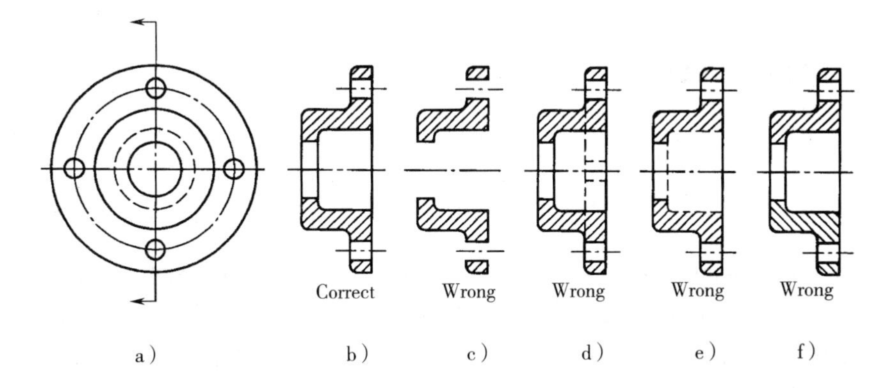

Figure 1-1-5 Lines in sectioning

A section-lined area is always completely bounded by a visible outline—never by a hidden line, as in Figure 1-1-5e, since in every case the cut surfaces and their boundary lines will be visible. Also, a visible line can never cut across a section-lined area.

In a sectional view of an object, alone or in assembly, the section lines in all sectioned areas must be parallel, not as shown in Figure 1-1-5f. The use of section lining in opposite directions is an indication of different parts, as when two or more parts are adjacent in an assembly drawing.

Unit 2 Tolerances

Text

Interchangeable manufacturing allows parts made in widely separated localities to be brought together for *assembly*. That the parts all fit together properly is an essential element of mass production. Without interchangeable manufacturing, modern industry could not exist, and without effective size control by the engineer, interchangeable manufacturing could not be achieved①.

However, it is impossible to make anything to exact size. Parts can be made to very close dimensions, even to a few millionths of an inch or thousandths of a *millimeter*, but such *accuracy* is extremely expensive.

Fortunately, exact sizes are not needed. The need is for varying degrees of accuracy according to functional requirements. A manufacturer of children's *tricycles* would soon go out of business if the parts were made with jet-engine accuracy—no one would be willing to pay the price②. So what is wanted is a means of specifying dimensions with whatever degree of accuracy is required. The answer to the problem is the specification of a *tolerance* on each dimension.

Tolerance is the total amount that a specific dimension is permitted to vary; it is the difference between the maximum and the minimum *limits* for the dimension. It can be specified in any of the two forms: *unilateral* or *bilateral*. In unilateral tolerance, the variation of the size will be *wholly* on the side. For example, $30^{\ 0}_{-0.02}$ is a unilateral tolerance. Here the nominal dimension 30 is allowed to vary between 30mm and 29.98mm. In bilateral tolerance, the variation will be to both the sides. For example, 30.00 ± 0.01 or $30^{+0.05}_{-0.10}$. In bilateral tolerance, the variation of the limits can be uniform as shown in the former case. The dimension varies from 30.01mm to 29.99mm. *Alternatively* the allowed *deviation* can be different as shown in the second case. Here the dimension varies from 30.05mm to 29.90mm.

In engineering when a product is designed it consists of a number of parts and these parts mate with each other in some form. In the assembly it is important to consider the type of mating or fit between two parts which will actually define the way the parts are to behave during the working of the assembly.

Take for example a shaft and hole, which will have to fit together. In the simplest case if the dimension of the shaft is lower than the dimension of the hole, then there will be *clearance*. Such a fit is termed clearance fit. Alternatively, if the dimension of the shaft is more than that of the hole, then it is termed interference fit. These are illustrated in Figure 1-2-1a and b. However in Figure 1-2-1c, depending upon the possibilities of dimensions, at times there will be clearance and other times there will be *interference*. Such a fit is termed as *transition* fit.

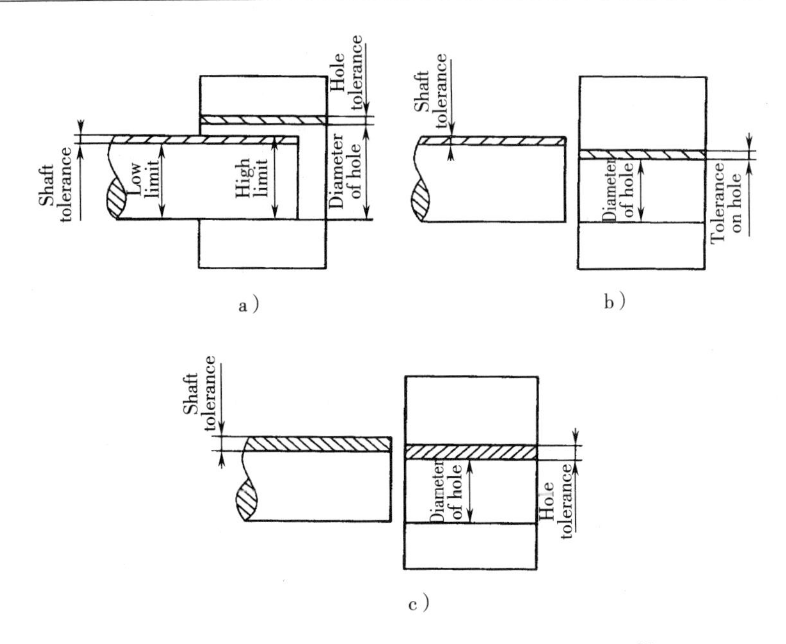

Figure 1-2-1 Typical fits possible in engineering assemblies

Questions

1. Why is it impossible to make anything to exact size?
2. What is the meaning of tolerance?
3. What is the difference between unilateral tolerance and bilateral tolerance?
4. Explain the concepts of clearance, interference and transition fits.

New Words and Expressions

1. interchangeable [intə'tʃeindʒəbl] adj. 可交换的，可互换的
2. interchangeability [intəːtʃeindʒə'biliti] n. 互换性
3. assembly [ə'sembli] n. 集合；装配，组件；装配图
4. millimeter ['miliˌmiːtə] n. 毫米
5. accuracy ['ækjurəsi] n. 准确（度），精确（度）
6. fortunate ['fɔːtʃənət] adj. 幸运的，侥幸的
7. tricycle ['traisikl] n. 三轮车
8. tolerance ['tɔlərəns] n. 公差，容差
9. limit ['limit] n. 界限，范围；极限 vt. 限制
10. unilateral [juːni'lætərəl] adj. 单边的，单向的
11. bilateral [bai'lætərəl] adj. 双边的，两边的
12. wholly ['həulli] adv. 完全地，实足
13. alternatively [ɔːl'təːnətiv] adj. 两者挑一的，交替的；选择的
14. deviation [diːvi'eiʃən] n. 偏差，偏离，偏向
15. clearance ['kliərəns] n. 间隙，空隙

16. interference [intəˈfiərəns] n. 过盈，干涉，抵触
17. transition [trænˈsiʃn] n. 过渡；转变，转移
18. as a means of 作为……的工具（或方法，手段）
19. be termed as 被叫做，被称作
20. in the former case 在前一种情形下

Notes

[1] Without interchangeable manufacturing, modern industry could not exist, and without effective size control by the engineer, interchangeable manufacturing could not be achieved.

没有互换性生产，现代工业就不可能存在；而没有工程师对零件尺寸的有效控制，互换性生产就不可能实现。

句中由介词 without 引导两个假设条件句，分别在句中做状语，without 作“没有”解。

[2] A manufacturer of children's tricycles would soon go out of business if the parts were made with jet-engine accuracy—no one would be willing to pay the price.

如果童车制造商将童车制造成与喷气式发动机一样的精度，这样不但没有人情愿支付昂贵的价格来购买，而且制造商还将面临无法经营的境地。

本句主体为由 if 引导的与现实事实相反的虚拟语气结构。

Glossary of Terms

1. unilateral tolerance 单边间隙
2. bilateral tolerance 双边间隙
3. clearance fit 间隙配合
4. interference fit 过盈（静）配合
5. transition fit 过渡配合
6. hole-basis (basic-hole) system 基孔制
7. shaft-basis (basic-shaft) system 基轴制
8. basic size 基本尺寸，公称尺寸
9. actual size 实际尺寸
10. limit of size 极限尺寸
11. upper (lower) derivation 上（下）偏差
12. error 误差
13. tolerance on fit 配合公差
14. tolerance zone 公差带
15. mass production 成批生产，大批生产
16. standard tolerance 标准公差
17. tolerance grade 公差等级
18. nominal error 名义误差
19. geometric tolerance 几何公差

20. positional tolerance 位置公差
21. working (finishing) allowance 加工余量
22. straightness, flatness, circularity, cylindricity, parallelism, perpendicularity 直线度，平面度，圆度，圆柱度，平行度，垂直度
23. angularity, concentricity, symmetry, roughness, finishing 倾斜度，同轴度，对称度，表面粗糙度，光洁度
24. total runout (runout) 全跳动（圆跳动）
25. datum (~ line, ~ plane) 基准（基准线，基准面）
26. setting up error 安装误差
27. dimensional control 尺寸控制
28. unilateral limit 单边极限
29. bilateral limit 双边极限

Reading Materials

Hole-Basis and Shaft-Basis System

For obtaining the required fit, the organization can choose any one of the following two possible systems.

Hole-basis system. In this system the nominal size and the limits on the hole are maintained constantly and the shaft limits are varied to obtain the requisite fit. For example,

Let the hole size be $30.00^{+0.03}_{0}$.

Shaft of $30.00^{+0.02}_{-0.01}$ gives the transition fit.

Shaft of $30.00^{+0.08}_{+0.04}$ gives the interference fit.

Shaft of $30.00^{-0.01}_{-0.04}$ gives the clearance fit.

Shaft-basis system. This is the reverse of hole-basis system. In this system the shaft size and limits are maintained constant while the limits of hole vary to obtain any fit.

Though there is not much to choose between the two systems, the hole-basis system is mostly used because standard tools such as reamers, drills, broaches and other standard tools are often used to produce holes, and standard plug gages are used to check the actual sizes. On the other hand, shafting can easily be machined to any size desired.

Preferred Fits

The symbols for either the hole-basis or shaft-basis preferred fits (clearance, transition, and interference) are given in Table 1-2-1. Fits should be selected from this table for mating parts where possible.

Although the second and the third choice basic size diameters are possible, they must be calculated from tables not included in this text. For the generally preferred hole-basis system, note that the ISO symbols range from H11/c11 (loose fit) to H7/u6 (force fit). For the shaft-basis system, the preferred symbols range from C11/h11 (loose fit) to U7/h6 (force fit).

Table 1-2-1 Preferred fits

Fits	ISO symbol		Description
	Hole-basis	Shaft-basis	
Clearance Fits	H11/c11	C11/h11	Loose-running fit for wide commercial tolerances or allowances on external members
	H9/d9	D9/h9	Free-running fit not for use where accuracy is essential, but good for large temperature variation, high running speeds, or heavy journal pressures
	H8/f7	F8/h7	Close-running fit for running on accurate machines and for accurate location at moderated speeds and journal pressures
Transition Fits	H7/g6	G7/h6	Sliding fit not intended to run freely, but to move and turn freely and locate accurately
	H7/h6	H7/h6	Locational clearance fit provides snug fit for locating stationary parts, but can be freely assembled and disassembled
	H7/k6	K7/h6	Locational transition fit for accurate location, a compromise between clearance and interference
	H7/n6	N7/h6	Locational transition fit for more accurate location where greater interference is permissible
Interference Fits	H7/p6	P7/h6	Locational interference fit for parts requiring rigidity and alignment with prime accuracy of location but without special bore pressure requirements
	H7/s6	S7/h6	Medium drive fit for ordinary steel parts or shrink fits on light sections, the tightest fit usable with cast iron
	H7/u6	U7/h6	Force fit suitable for parts which can be highly stressed or for shrink fits where the heavy pressing forces required are impractical

Symbols for Tolerances of Position and Form

Since traditional narrative notes for specifying tolerance of position (location) and form (shape) may be confusing or unclear, may require too much space, and may not be understood internationally, most multinational companies have adopted symbols for such specifications (ANSI/ASME Y14.5M—1994). These ANSI symbols provide an accurate and concise means of specifying geometric characteristics and tolerances in a minimum of space (Table 1-2-2). The symbols may be supplemented by notes if the precise geometric requirements cannot be conveyed by the symbols.

Table 1-2-2 Tolerance symbols of form, orientation and location

Geometric characteristic symbols				Modifying symbols	
	Type of tolerance	**Characteristic**	**Symbol**	**Term**	**Symbol**
For individual features	Form	Straightness	—	At maximum material condition	Ⓜ
		Flatness	⏥	At least material condition	Ⓛ
		Circularity (roundness)	○	Projected tolerance zone	Ⓟ
		Cylindricity	⌭	Free state	Ⓕ
For individual or related features	Profile	Profile of a line	⌒	Tangent plane	Ⓣ
		Profile of a surface	⌓	Diameter	ϕ
For related features	Orientation	Angularity	∠	Spherical diameter	$R\phi$
		Perpendicularity	⊥	Radius	R
		Parallelism	//	Spherical radius	SR
	Location	Position	⌖	Controlled radius	CR
		Concentricity	◎	Reference	( )
		Symmetry	⌯	Arc length	⌒
	Runout	Circular runout	↗	Statistical tolerance	⟨ST⟩
		Total runout	⌰	Between	↔

Combinations of the various symbols and their meanings are given in Figure 1-2-2.

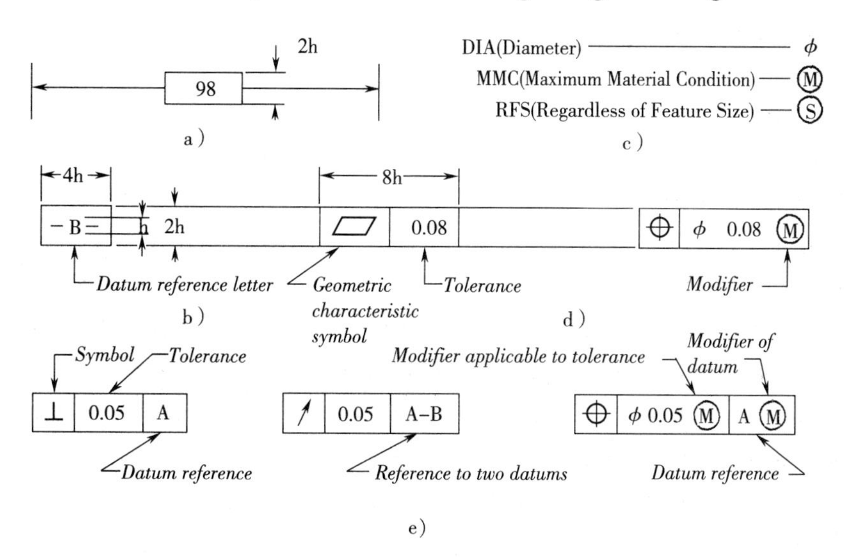

Figure 1-2-2 Use of symbols for tolerance of position and form (ANSI/ASME Y14. 5M—1994)

a) Basic dimension symbol b) Datum symbol c) Modifying symbols

d) Feature control symbols e) Feature control symbols with datum references

Unit 3 Manufacturing Processes

Text

There are a large number of *processes* available for *manufacture* to the engineer. These processes can be broadly classified into four categories:

1. *Casting* or molding processes
2. Forming or metalworking processes
3. *Fabrication* processes
4. Material removal (machining) processes

Casting processes. These are the only processes where liquid metal is used. Casting is also the oldest known manufacturing process. It requires a cavity usually in a refractory material to resemble closely the final object to be made. Molten metal is poured into this refractory mould cavity and allowed to solidify. The object after *solidification* will be removed from the mould. Casting processes are universally used to manufacture a wide variety of products. The principal process among these is sand casting where sand is used as the refractory material. The process is equally suitable for the production of a small batch as well as a large scale.

Some of the other casting processes for specialized needs are:

(1) Shell mould casting

(2) Precision investment casting

(3) *Plaster* mould casting

(4) Permanent mould casting

(5) Die casting

(6) *Centrifugal* casting

Forming processes. These are solid state manufacturing processes involving minimum amount of material wastage and faster production. In a forming process, metal may be heated to a temperature which is slightly below the solidus temperature and then a large force is applied such that the material flows and takes the desired shape①. The desired shape is controlled by means of a set of tools called dies which may be completely closed or partially closed during manufacture. These processes are normally used for large scale production . These processes are generally economical and in many cases improve the mechanical properties too.

Some of the metal forming processes are:

(1) *Rolling*

(2) Drop *forging*

(3) Press forging

(4) Upset forging

(5) Extrusion

(6) Wire drawing

(7) Sheet metal operations

Fabrication processes. These are secondary manufacturing processes where the raw materials are processed by any of the manufacturing processes previously described. It essentially involves joining pieces either permanently or temporarily so that they would perform the necessary function. The joining can be achieved by either or both of heat and pressure with a joining material. Many of the steel structural constructions we see are first rolled and then joined together by a fabrication process.

Some of the processes of interest in this category are:

(1) Gas *welding*

(2) Electric arc welding

(3) Electric resistance welding

(4) Cold welding

(5) *Brazing*

(6) *Soldering*

(7) Mechanical *fastening*

(8) Adhesive bonding

Material removal processes. These are also the secondary manufacturing processes where the additional unwanted material is removed in the form of chips from the blank material by a harder tool so as to obtain the final desired shape. Material removal is normally the most expensive manufacturing process because more energy is consumed, and also a lot of waste material is generated in the process②. Still this is widely used because it delivers very good dimensional accuracy and good surface finish, it also generates accurate contours. Material removal processes are also called machining processes.

Various processes in this category are:

(1) *Turning*

(2) *Drilling*

(3) Shaping and planning

(4) *Milling*

(5) *Grinding*

(6) *Broaching*

(7) Sawing

All these manufacturing processes have been continuously developed so as to obtain better products at a reduced cost. Of particular interest is the development of computers and their effect on the manufacturing processes. The advent of computers has made a remarkable difference to most of the above manufacturing processes. They have contributed greatly to both automation and designing of the processes.

Questions

1. What are the basic manufacturing processes?
2. Why is it necessary for all engineers to be familiar with manufacturing processes?
3. What are the forming processes of engineering materials?
4. What are the material removal processes of the metals?

New Words and Expressions

1. process [ˈprəuses] n. 过程，工序，工艺规程 adj. 经加工的，有特殊光效的；vt. 处理，加工
2. manufacture [ˌmænjuˈfæktʃə] vt. （大量）制造，加工 n. （大量）制造，制造业
3. casting [ˈkɑːstiŋ] n. 铸造，铸件
4. fabrication [ˌfæbriˈkeiʃn] n. 制配，生产，建造
5. solidification [səˌlidifiˈkeiʃn] n. 凝固
6. plaster [ˈplɑːstə] n. 石膏
7. centrifugal [senˈtrifjəgl] adj. 离心力的 n. 离心式压缩机
8. rolling [ˈrəuliŋ] n. 轧制，辗压 adj. 滚动的，旋转的
9. forging [ˈfɔːdʒiŋ] n. 锻造，锻件 adj. 锻造的
10. welding [ˈweldiŋ] n. 焊接，熔接 adj. 焊的
11. brazing [ˈbreiziŋ] n. 铜焊，硬焊
12. soldering [ˈsəuldəriŋ] n. 软焊，锡焊，钎焊
13. fastening [ˈfɑːsniŋ] n. & adj. 连接（的），坚固件，固定（的）
14. turning [ˈtəːniŋ] n. 旋转，车削，外圆切削，切屑
15. drilling [ˈdriliŋ] n. 钻孔，穿孔
16. milling [ˈmiliŋ] n. 铣削，铣（法）；研磨
17. grinding [ˈgraindiŋ] n. 磨，磨削 adj. 磨的
18. broach [brəutʃ] n. 拉刀，扩孔器 vt. 扩孔，拉削
19. in process 在进行中
20. in turn 依次，轮流
21. a small batch 小批量
22. a large scale 大规模

Notes

[1] In a forming process, metal may be heated to a temperature which is slightly below the solidus temperature and then a large force is applied such that the material flows and takes the desired shape.

在成形工艺过程中，可以先将金属材料加热到略低于固相线的温度，然后给金属材料施加较大的压力，从而使材料流动并形成所需要的形状。

在句中，which is slightly below the solidus temperature 为限制性定语从句，修饰 a temperature.

[2] Material removal is normally the most expensive manufacturing process because more energy is consumed, and also a lot of waste material is generated in the process.

材料去除加工通常是最为昂贵的制造工艺，因为在这种工艺过程中消耗的能量更多，并且还会产生许多废料。

句中 because 作“因为”、“由于”解，连接原因状语从句；a lot of (much)，可译为“许多”。

Glossary of Terms

1. casting or molding processes 铸造成形工艺
2. forming or metalworking processes 金属成形工艺
3. fabrication processes 金属制造工艺
4. material removal (machining) processes 金属去除（机械加工）工艺
5. shell mould casting 壳型铸造
6. precision investment casting 精密铸造
7. plaster mould casting 石膏型铸造
8. permanent mould casting 永久型铸造
9. die casting 压力铸造
10. centrifugal casting 离心铸造
11. rolling 轧制
12. drop forging 落锤锻造
13. press forging 压力锻造
14. forging shop 锻工车间
15. upset forging 镦锻
16. extrusion 挤压
17. wire drawing 拉丝，拔丝
18. sheet metal operations 金属板料加工
19. gas welding 气焊
20. electric arc welding 电弧焊
21. electric resistance welding 电阻焊
22. cold welding 冷焊
23. soldering iron 烙铁，焊铁
24. mechanical fastening 机械连接（固定）
25. rolling friction 滚动摩擦
26. refractory material 耐火材料

Reading Materials

Typical Comparison of Different Manufacturing Processes

As mentioned earlier, material removal processes are very expensive and hence should be resorted to only when absolutely required. Table 1-3-1 gives a relative comparison of material removal processes with other manufacturing processes.

Table 1-3-1 Typical comparison of different manufacturing processes

Manufacturing process	Typical application	Size range	Tolerance, surface finish
Sand casting	All metals	Unlimited	±0.03mm; 3.2μm
Die casting	Zinc and aluminum alloys	Up to 10 kg	±0.015mm; 1.6μm
Drop forging	All metals	Unlimited	—
Hot extrusion	All metals	Unlimited	—
Gas metal arc welding	All metals	12mm thick	—
Sheet metal blanking	All materials	—	±0.08mm
Turning	All materials	Unlimited	±0.05mm; 2.0μm
Milling	All materials	Unlimited	±0.05mm; 2.0μm
Grinding	All materials	Unlimited	±0.025mm; 0.4μm
Electric discharge machining	Electrically conductive materials	—	±0.003mm; 0.1μm

Selecting Manufacturing Processes

An extensive and continuously expanding variety of manufacturing processes are used to produce parts and there is usually more than one method of manufacturing a part for the given material. The broad categories of the processing methods for materials are as follows, referenced to the relevant part in this text and illustrated with examples for each:

Casting: expendable molding and permanent molding.

Forming and shaping: rolling, forging, extrusion, drawing, sheet forming, powder metallurgy and molding.

Machining: turning, boring, drilling, milling, planning, shaping, broaching, grinding, and ultrasonic machining; chemical, electrical and electrochemical machining; high-energy-beam machining; and this category also includes micromachining for ultraprecision parts producing.

Joining: welding, brazing, soldering, diffusion bonding, adhesive bonding and mechanical joining.

Finishing: honing, lapping, polishing, burnishing, deburring, surface treating, coating and plating.

Nanofabrication: it is the most advanced technology and is capable of producing parts with dimensions at the nano level (one billionth of a meter); it typically involves processes such as etching techniques, electron-beams and laser-beams. Present applications are in the fabrication of microelectromechanical systems and extending to nanoelectromechanical systems, which operate on the same scale as biological molecules.

Selection of a particular manufacturing process, or a sequence of processes, depends not only on the shape to be produced but also on other factors pertaining to the material properties. For examples, brittle and hard materials cannot be shaped or formed easily, whereas they can be cast, machined, or ground. Metals that are previously formed at room temperature become stronger and less ductile than they were before being processed, and thus it will require higher forces and be less formable during subsequent processing.

As described throughout this text, each manufacturing process has its own advantages and limitations, as well as production rates and production costs. Manufacturing engineers are constantly being challenged to find new solutions to manufacturing problems and cost reduction. For example, sheet metal parts traditionally have been cut and fabricated using common mechanical tools, punches and dies. Although still widely used, some of these operations are being replaced by laser-cutting techniques in which the path of the laser can be controlled, thus increasing the capability, as well as eliminating the need of punches and dies. However, as expected, the surface produced by punching has different characteristics from that produced by laser-cutting.

Unit 4 Properties of Engineering Materials

Text

Design of structures and systems requires determination of component dimensions and is based on the appropriate mechanical properties of materials.

Tension testing. The *tension* testing is the test most commonly used to evaluate the mechanical properties of materials. A typical load-*elongation* curve for a pure metal is shown in Figure 1-4-1. A number of important quantities can be calculated from the load-elongation or stress-strain curve of a material, namely:

(1) ***Modulus* of elasticity**. The modulus of elasticity, or Young's modulus (E), is defined as the tensile stress divided by the tensile strain for elastic deformation and so is the *slope* of the linear part of the stress-strain curve. This relationship is Hooke's Law: $E = \sigma/\varepsilon$.

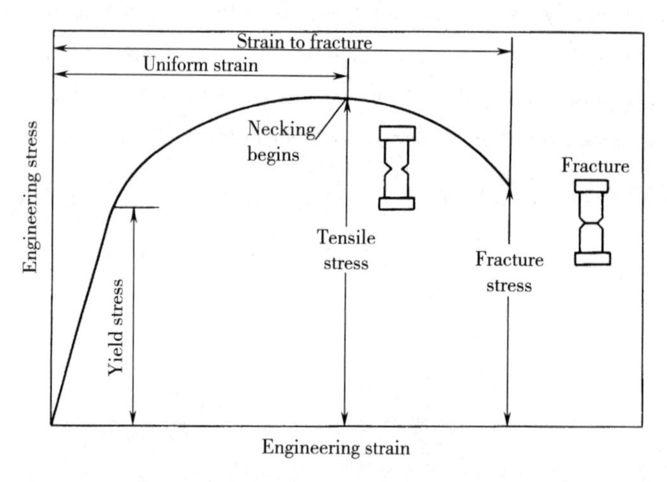

Figure 1-4-1 Engineering stress-strain curve

(2) **Yield strength**. When a material under tension reaches the limit of its elastic strain and begins to flow plastically, it is said to have yielded①. The yield strength is then the stress at which plastic flow starts.

(3) **Tensile strength**. This is defined as the maximum load *sustained* by the specimen during the tensile test, divided by its original cross-sectional area. It is sometimes called the *ultimate* strength of the material.

(4) **Tensile elongation**. This is frequently taken as an indicator of the *ductility* of the material under tensile test. To determine the elongation, the increase in distance between two reference marks, *scribed* on the specimen before test, is measured with the two halves of the broken specimen held together②. The percentage elongation is 100 times the quotient of the increase in length and the initial distance between the scribed marks.

(5) **Reduction of area**. This is the *quotient* of the decrease in cross-sectional area at the plane of fracture and the original area at that plane (times 100, to express as a percentage). Similarly to percentage elongation, this number is related to ductility.

The shape and magnitude of the stress-strain curve of a metal will depend on its composition, heat treatment, prior history of plastic deformation, the strain rate, temperature and state of stress imposed during the testing. The parameters, which are used to describe the stress-strain curve of a metal, are the tensile strength, yield strength or yield point, percent elongation and reduction of area. The first two are strength parameters; the last two indicate ductility.

Questions

1. What can be calculated from the long-elongation or stress-strain curve?
2. What is the tensile strength?
3. What is ductility? How to determine the elongation?
4. What is yield strength?

New Words and Expressions

1. tension [ˈtenʃn] n. & v. 拉伸，拉紧
2. elongation [ˌiːlɔŋˈgeiʃn] n. 伸长，伸长率
3. modulus [ˈmɔdjuləs] n. 模量，系数，率
4. slope [ˈsləup] n. 倾斜面，斜度，坡（梯）度
5. sustain [səsˈtein] vt. 持续，支撑
6. ultimate [ˈʌltimət] n. 极限，终极，极点
7. ductility [dʌkˈtiliti] n. 韧性，塑性
8. scribe [skraib] n. 划线器 vt. 用划线器划线
9. quotient [ˈkwəuʃnt] n. 商数，系数
10. load-elongation curve 力-伸长曲线
11. define... as... 把……定义为；把……规定为
12. be used to do 用于做……

Notes

[1] When a material under tension reaches the limit of its elastic strain and begins to flow plastically, it is said to have yielded.

当材料在拉力作用下达到它的弹性应变极限而开始产生塑性流变时，这种现象叫作屈服。

句中 it 做形式主语，when 引导的状语从句做真正的主语。

[2] To determine the elongation, the increase in distance between two reference marks, scribed on the specimen before test, is measured with the two halves of the broken specimen held together.

为了确定伸长量，即试样上两个参考标记间距的增加量，拉伸试验前先在试样上划线作为标记，拉断后将两部分试样放在一起进行测量。

在本句中，the increase in distance between two reference marks 做 the elongation 的同位语；而 scribed on the specimen before test 则做 marks 的定语。

Glossary of Terms

1. mechanical properties　力学性能（机械性能）
2. physical properties　物理性能
3. chemical properties　化学性能
4. technique properties　工艺性能
5. cross-sectional area　横断面
6. impact-loading　冲击载荷
7. fatigue and endurance limit　疲劳极限，持久极限
8. engineering stress　工程应力
9. engineering strain　工程应变
10. elastic deformation　弹性变形
11. elastic limit　弹性极限
12. proportional limit　比例极限
13. modulus of elasticity　弹性模量
14. stiffness　刚度
15. yield strength　屈服强度
16. tensile strength　抗拉强度
17. impact toughness　冲击韧度
18. ductility and brittleness　塑性和脆性
19. Rockwell hardness testing　洛氏硬度试验
20. Brinell hardness testing　布氏硬度试验
21. Vickers hardness testing　显微硬度试验
22. stress-strain curve　应力-应变曲线
23. brittle material　脆性材料
24. fracture surface　断裂面
25. toughness　坚韧性，刚性
26. fatigue strength　疲劳强度
27. melting point　熔点
28. thermal expansion　热膨胀
29. thermal conductivity　导热性
30. hot hardness　热硬性

Reading Materials

Compression Testing

Compression testing is an extremely valuable testing procedure that is often overlooked because it is not properly understood. One of the main advantages of the compression test is that tests can be performed with a minimum of material, and thus mechanical properties can be obtained from specimens that are too small for tension testing. Compression tests are also very helpful for predicting the bulk formability of materials (behavior in forging, extrusion, rolling, etc.).

In compression testing, the material does not neck as in tension, but undergoes barreling; failure occurs by different mechanisms and therefore there is no UTS (ultimate tensile strength). In general, ductile materials do not fail in compression but tend to flow in response to the imposed loads. Brittle cylindrical specimens loaded in compression fail in shear on a plane inclined to the load, and therefore actually break into two or more pieces. In this case, an ultimate (compressive) stress can be defined.

In comparison with tension testing, several difficulties are encountered in conducting compression tests and interpretation of the experimental data. For example, maintaining complete axiality of the applied load is important. In tension testing, self-aligning grips make this relatively simple to accomplish. In compression testing, if the specimen is tall in relation to its diameter, this can present a major difficulty.

Hardness Testing

The hardness testing measures the resistance to penetration of the surface of a material by a hard object. A variety of hardness tests have been devised, but the most commonly used are the Rockwell testing and the Brinell testing.

In the Brinell hardness testing a hard steel sphere, usually 10mm in diameter, is forced into the surface of the material. The diameter of the impression left on the surface is measured and the Brinell hardness number (BHN) is calculated from the following equation:

$$\text{BHN} = \frac{F}{(\pi/2)D(D - \sqrt{D^2 - D_t^2})}$$

Where F is the applied load in kilograms, D is the diameter of the indentor in millimeters, and D_t is the diameter of the impression in millimeters.

The Rockwell hardness testing uses either a small diameter steel ball for soft materials or a diamond cone, or Brale, for harder materials. The depth of penetration of the indentor is automatically measured by the testing machine and converted to a Rockwell hardness number.

The Vickers and Knoop testings are microhardness testings; they form such small indentations that a microscope is required to obtain the measurement.

The hardness numbers are used primarily as a basis for comparison of materials, specifications for manufacturing and heat treatment, quality control, correlation with other properties and behavior of materials. For example, Brinell hardness is closely related to the tensile strength of steel by the relationship

$$\text{Tensile strength} = 500\ \text{BHN}$$

A Brinell hardness number can be obtained in just a few minutes with virtually no preparation of the specimen and without destroying the component, yet provides a close approximation for the tensile strength.

Hardness correlates well with wear resistance. A material used to crush or grind ore should be very hard to assure that the material is not eroded or abraded by the hard feed materials. Similarly, gear teeth in a transmission or drive system of a vehicle should be hard so that the teeth do not wear out.

Impact Testing

A material may have a high tensile strength and yet be unsuitable for shock loading conditions. To determine this, the impact resistance is usually measured by means of the notched or unnotched Izod or Charpy impact test. In this test a load swings from a given height to strike the specimen, and the energy dissipated in the fracture is measured. The energy absorption is supposed to reveal the toughness of metals under impact-loading conditions.

Fatigue Testing

Fatigue testing determines the ability of a material to withstand repeated applications of a stress, which is too small to produce appreciable plastic deformation. Several different types of testing machines have been constructed in which the stress is applied by bending, torsion, tension or compression, but all involve the same principle of subjecting the material to constant cycles of stress. To express the characteristics of the stress system, three properties are usually quoted, these include: ① the maximum range of stress, ② the mean stress, and ③ the time period for the stress cycle. The standard method of studying fatigue is to prepare a large number of specimens free from flaws and to subject them to tests using a different range of stress, S, on each group of specimens. The number of stress cycles, N, endured by each specimen at a given stress level is recorded and plotted. This S-N diagram indicates that some metals can withstand indefinitely the application of a large number of stress reversals, provided the applied stress is below a limiting stress known as the endurance limit.

Chapter 2 Engineering Materials and Heat Treatment

Unit 1 Engineering Materials

Text

There are different ways of classifying materials. One way is to divide them into five groups: metals and alloys; *ceramics*; *polymers* (plastics); *composite* materials; and *semiconductors*. Materials in each of these groups possess different structures and properties.

Metals and alloys. These include steel, aluminum, magnesium, zinc, cast iron, titanium, copper and nickel. In general, metals have good electrical and thermal conductivity. Metals and alloys have relatively high strength, high stiffness, ductility or formability, and shock resistance. They are particularly useful for structural or load-bearing applications. Although pure metals are occasionally used, combinations of metals called alloys provide improvement in a particular desirable property or permit better combinations.

Ceramics. Ceramics can be defined as inorganic crystalline materials. Ceramics are probably the most "natural" materials. Beach sand and rocks are examples of naturally occurring ceramics. Advanced ceramics are materials made by refining naturally occurring ceramics and other special processes. Advanced ceramics are used in substrates that of house computer chips, sensors and *actuators*, capacitors, wireless communications, spark plugs, inductors and electrical insulation. Some ceramics are used as barrier coatings to protect metallic substrates in turbine engines. Ceramics are also used in such consumer products as paints, plastics and tires, and for industrial applications such as the tiles for the space *shuttle*, a *catalyst* support, and the oxygen sensors used in cars①. Traditional ceramics are used to make bricks, tableware, *sanitaryware*, *refractories* (heat-resistant materials) and abrasives. In general, due to the presence of porosity (small holes), ceramics do not conduct heat well and must be heated to very high temperatures before melting. Ceramics are strong and hard, but also very brittle. We normally prepare fine powders of ceramics and convert them into different shapes. New processing techniques make ceramics sufficiently resistant to fracture that they can be used in load-bearing applications, such as *impellers* in turbine engines.

Polymers (plastics). Polymers are typically organic materials. They are produced by using a process known as polymerization. Polymeric materials include rubber (elastomer) and many types of adhesives. Many polymers have very good electrical resistivity. They can also provide good thermal insulation. Although they have lower strength, polymers have a very good strength-to-weight ratio. They are typically not suitable for use at high temperatures. Many polymers have very good resistance to corrosive chemicals. Polymers have thousands of applications ranging from *bulletproof* vests, compact disks (CDs), ropes and liquid crystals (LCDs) to clothes and coffee cups②. Thermoplastic polymers, in which the long molecular chains are not rigidly connected, have good ductility and formability. Thermosetting polymers are stronger but more brittle because the molecular chains are tightly linked. Polymers are used in many applications, including electronic devices. Thermoplastics are made by shaping their molten form. Thermosets are typically cast in molds. The term plastic is used to describe polymeric materials containing additives.

Composite materials. The main idea in developing composites is to blend the properties of different materials. These are formed from two or more materials, producing properties not found in any single material. *Concrete*, *plywood* and fiberglass are examples of composite materials. Fiberglass is made by dispersing glass fibers in a polymer matrix. The glass fibers make the polymer stiffer, without significantly increasing its density. With composites we can produce lightweight, strong, ductile, high temperature-resistant materials or we can produce hard, yet shock-resistant, cutting tools that would otherwise *shatter*. Advanced aircraft and aerospace vehicles rely heavily on composites such as carbon-fiber-reinforced polymers. Sports equipment such as bicycles, golf clubs, tennis rackets and the like also make use of different kinds of composite materials that are light and stiff.

Semiconductors. Silicon, germanium and gallium arsenide-based semiconductors such as those used in computers and electronics are part of a broader class of materials known as electronic materials. The electrical conductivity of semiconducting materials is between that of ceramic insulators and metallic conductors. Semiconductors have enabled the information age. In some semiconductors, the level of conductivity can be controlled to enable their use in electronic devices such as transistors, diodes, etc., which are used to build integrated circuits. In many applications, we need large single crystals of semiconductors. These are grown from molten materials. Often, thin films of semiconducting materials are also made by using specialized processes.

Questions

1. What are engineering materials?
2. Explain the properties of metals and alloys.
3. What are spark plugs made from?
4. What are bulletproof vests made from?
5. What is a composite? What do the properties of composite materials depend upon?

New Words and Expressions

1. ceramic [sə'ræmik] adj. 陶瓷的，陶器的 n. (pl.) 陶瓷制品
2. polymer ['pɔlimə(r)] n. 聚合物，多聚物
3. composite ['kɔmpəzit] adj. 合成的，组合的 n. 复合材料

4. semiconductor [ˌsemikənˈdʌktə (r)] n. 半导体
5. actuator [ˈæktjueitə] n. 传动装置，传动机构
6. shuttle [ˈʃʌtl] n. 穿梭，航天飞机
7. catalyst [ˈkætəlist] n. 接触剂，催化剂
8. sanitary [ˈsænitəri] adj. 环境卫生的，保持清洁的
9. refractory [riˈfræktəri] adj. 耐火的 n. 耐火材料
10. impeller [imˈpelə] n. 推进器；转子，叶轮
11. bulletproof [ˈbulitpruːf] adj. 防弹的
12. concrete [ˈkɔnkriːt] n. 混凝土 adj. 固结的 vt. 使凝固
13. plywood [ˈplaiwud] n. 胶合板，层板
14. shatter [ˈʃætə] v. 粉碎，损坏 n. 碎片
15. vest [vest] n. 汗衫 vi. 归属 vt. 授予
16. due to 由于，应归于
17. be resistant to 抵抗……的，防……的

Notes

[1] Ceramics are also used in such consumer products as paints, plastics and tires, and for industrial applications such as the tiles for the space shuttle, a catalyst support, and the oxygen sensors used in cars.

陶瓷像涂料、塑料和轮胎等消耗品一样也被用作消耗性产品，在工业应用方面可用于航天飞机瓷片、催化剂内衬和汽车上的氧传感器。

句中 such consumer products as paints..., such 与 as 分开，as 后为一个省略了 are 的定语从句，译为“像涂料、塑料和轮胎等消耗品一样”；句中 industrial applications such as... 为“复数名词 + such as... ”结构，such as 是复合连接词，引出同位语，可译为“例如”。

[2] Polymers have thousands of applications ranging from bulletproof vests, compact disks (CDs), ropes and liquid crystals (LCDs) to clothes and coffee cups.

高聚物的用途有成千上万，其应用范围可以从制作防弹背心、高密度磁盘、绳索、液晶到制作衣服和咖啡杯子。

句中 ranging from... to 可译为“范围从……到……”。

Glossary of Terms

1. materials engineering 材料工程
2. materials science 材料科学
3. materials science and engineering 材料科学与工程
4. polymeric material 高分子材料
5. organic material 有机材料
6. inorganic material 无机材料
7. crystalline material 晶体材料

8. amorphous material 非晶体材料
9. composite material 复合材料
10. metal-matrix composite（MMC） 金属基复合材料
11. ceramic-matrix composite（CMC） 陶瓷基复合材料
12. polymer-matrix composite（PMC） 聚合物基复合材料
13. reinforced concrete 钢筋混凝土
14. semi-conducting material 半导体材料
15. metallic material 金属材料
16. nonmetallic material 非金属材料
17. traditional ceramic 普通陶瓷
18. synthetic fiber 合成纤维
19. synthetic resin 合成树脂
20. synthetic rubber 合成橡胶
21. foamed plastic 泡沫塑料
22. glass fiber 玻璃纤维
23. thermoplastic 热塑性塑料
24. thermosetting plastic 热固性塑料
25. reinforced plastic 增强塑料
26. fiber-reinforced plastic 纤维增强塑料
27. properties of material 材料的性能
28. fire-resistive material 耐火材料
29. refractory brick 耐火砖
30. spark plug 火花塞

Reading Materials

Plastics

Plastic covers a range of synthetic or semi-synthetic polymerization products. They are composed of organic condensation or addition polymers and may contain other substances to improve performance or economics. There are few natural polymers generally considered to be "plastics". Plastics can be formed into objects or films or fibers. Their name is derived from the fact that many are malleable, having the property of plasticity. Plastics are designed with immense variation in properties such as heat tolerance, hardness, resiliency and many others. Combined with this adaptability, the general uniformity of composition and light weight of plastics ensure their use in almost all industrial segments.

Plastic may also refer to any material characterized by deformation or failure under shear stress; see plasticity and ductility.

Plastic can be classified in many ways but most commonly by their polymer backbone (polyvinyl chloride, polyethylene, acrylic, silicone, urethane, etc.). Other classifications include thermoplastics, thermoset, elastomer, engineering plastic, addition or condensation.

Many plastics are partially crystalline and partially amorphous in molecular structure, giving them both a melting point (the temperature at which the covalent bonds dissolve) and one or more glass transitions (temperatures at which the degree of cross-linking is substantially reduced).

Thermoplastics generally have a low thermal conductivity, a low elastic modulus, and are thermally softening. Consequently, machining them requires sharp tools with positive rake angles (to reduce cutting forces), large relief angles, small depths of cut and feed, relatively high speeds and proper support of the workpiece. External cooling of the cutting zone may be necessary to keep the chips from becoming gummy and sticking to the tools. Cooling usually can be achieved with a jet of air, vapor mist or water-soluble oils.

Thermosetting plastics are brittle and sensitive to thermal gradients during cutting; machining conditions generally are similar to those of thermoplastics.

Properties of Materials

When selecting materials for products, we firstly consider their mechanical properties: strength, toughness, ductility, hardness, elasticity, fatigue and creep. The mechanical properties specified for a product and its components should, of course, be appropriate to the conditions under which the product is expected to function. Next the physical properties of materials are considered: density, specific heat, thermal expansion and conductivity, melting point, electrical and magnetic properties. A combination of mechanical and physical properties is the strength-to-weight and stiffness-to-weight ratios of materials, particularly important for aerospace and automotive applications, as well as for sports equipment. Aluminum, titanium and reinforced plastics, for example, generally have higher such ratios than steels and cast irons.

Chemical properties also play a significant role, both in hostile and in normal environments. Oxidation, corrosion, general degradation of properties, toxicity and flammability of materials are among the important factors to be considered. In some commercial airline accidents, for example, many deaths have been caused by toxic fumes from burning nonmetallic materials in the aircraft cabin.

The manufacturing properties of materials determine whether they can be cast, formed, machined, joined and heat-treated with relative ease (Table 2-1-1). The method or methods used to process materials to the desired final shapes can affect the product's performance, service life and cost.

Table 2-1-1 General manufacturing characteristics of various alloys

Alloy	Aluminum	Copper	Gray cast iron	White cast iron	Nickel	Steels	Zinc
Castability	E	G or F	E	G	F	F	E
Weldability	F	F	D	VP	F	E	D
Machinability	E or G	G or F	G	VP	F	F	E

Note: E—excellent; G—good; F—fair; D—difficult; VP—very poor.

Unit 2 Cast Irons (Gray Cast Irons)

Text

In order to understand the *fabricating* characteristics of cast irons, it is necessary to become familiar with the characteristics of the metal and the various types and classifications that are available①.

The dividing point between "steels" and "cast irons" is the point where the percentage of carbon is 2.11%, where the eutectic reaction becomes possible. For steels, we concentrate on the eutectoid portion of the diagram (Figure 2-2-1) in which the solubility lines and the eutectoid isotherm are specially identified. The A_3 shows the temperature at which ferrite starts to form on cooling; the A_{cm} shows the temperature at which cementite starts to form; and the A_1 is the eutectoid temperature line.

One of the *distinguishing features* of all cast irons is that they have a relatively high carbon content. Steels range up to about 2% carbon. Cast irons overlap with the steels somewhat and range from about 1.5% up to 5% carbon. It is principally the form of the carbon, which is governed by thermal conditions and alloying elements, that provides various structures that may be classified into the following main types: gray cast iron; white cast iron; *malleable* iron; *ductile* (*nodular*) graphite iron; compacted (vermicular) graphite iron.

Gray cast iron. The terms gray and white cast iron refer to the appearance of the *fractured* area. The gray cast iron has a grayish appearance because of the larger amount of flake graphite on the surface. The dark sections show the graphite flakes. The pearlite type may be made fine by faster cooling, or *coarse* by slow cooling. The size of the section will also determine the structure of the metal; the thinner the section the faster it cools. Thus a casting having large variations in section will also have large variations in hardness and strength unless special *precautions* are taken to ensure uniform cooling.

The basic composition of gray cast irons is often described in terms of carbon *equivalent* (CE). This factor gives the relationship of the percentage of carbon and silicon in the iron to its capacity to produce graphite. Thus

$$CE = C_t + (1/3)(Si\% + P\%)$$

Where: C_t = total percentage of carbon.

The CE value may then be related to the tensile strength of the metal. Irons with a carbon equivalent of over 4.3% are called *hypereutectic* and are particularly good for thermal-shock resistance, such as for ingot molds②. The higher strength gray irons, with CE less than 4.3%, are termed *hypoeutectic*.

The compressive strength of gray cast iron is one of its *outstanding* features. In general, it ranges from 3 to 5 times the tensile strength. As an example, a class-20 gray cast iron which has a tensile strength of 20,000 psi has a compressive strength of 83,000 psi.

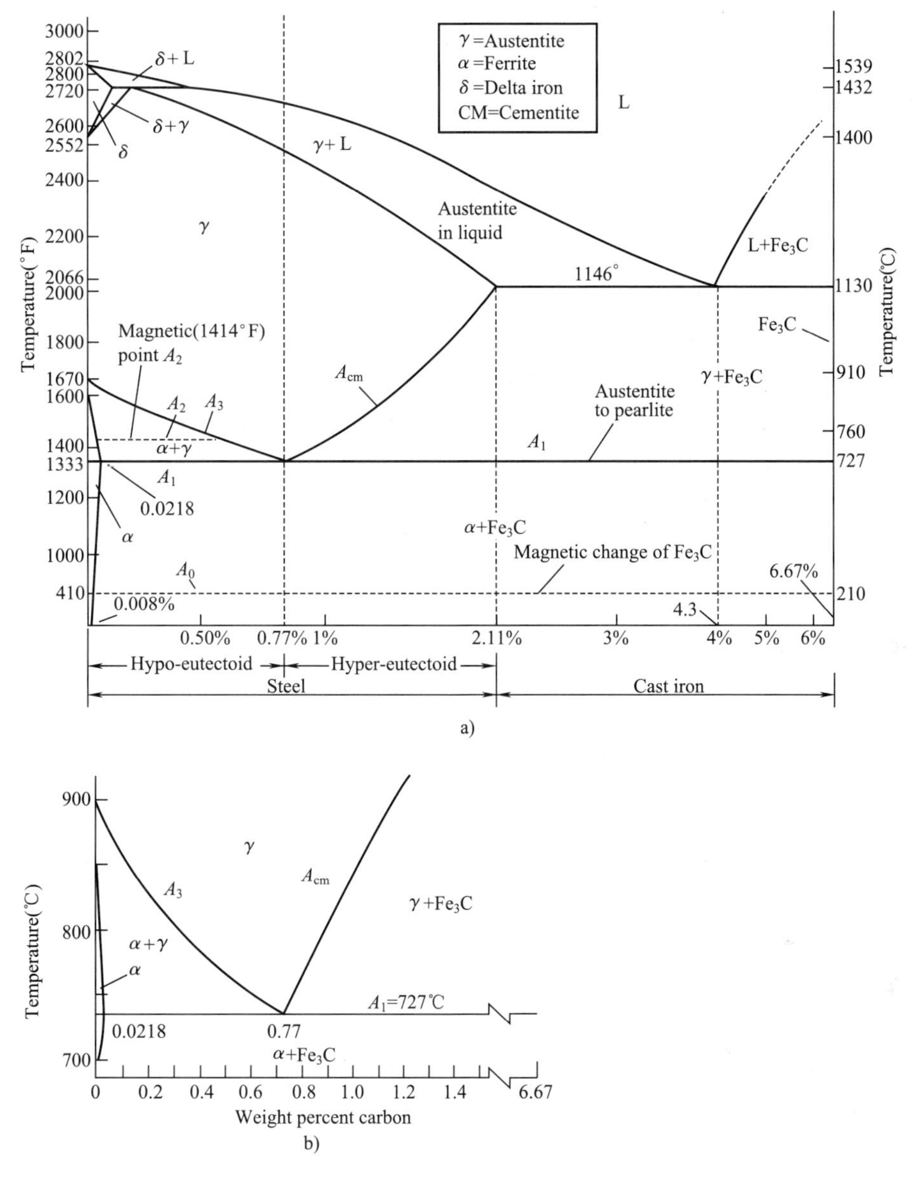

Figure 2-2-1

a) An expanded version of the Fe-C diagram.

b) The eutectoid portion of the Fe-Fe_3C phase diagram.

Gray cast iron is the most widely used of all cast metals. Typical applications are engine blocks, pipes and fittings, agricultural implements, bathtubs, household appliances, electric motor housings, machine tools, etc.

Questions

1. Describe the microstructure of gray cast irons.
2. What structural unit is generally altered to increase the strength of gray cast irons?
3. What are some of the attractive engineering properties of gray cast irons?

4. What is a carbon equivalent?

New Words and Expressions

1. fabricate [ˈfæbrikeit] vt. 制作，构成
2. distinguishing [disˈtiŋgwiʃiŋ] adj. 有区别的，明显的，有特征的
3. feature [ˈfiːtʃə (r)] n. 特征，特色，性能
4. malleable [ˈmæliəbl] adj. 有延展性的，可锻的
5. ductile [ˈdʌktail] adj. 易延展的
6. nodular [ˈnɔdjulə (r)] adj. 球状的，粒状的
7. fracture [ˈfræktʃə] n. 断口，断裂 v. （使）破碎
8. coarse [kɔːs] adj. 粗糙的
9. precaution [priˈkɔːʃn] n. 预防，警惕
10. equivalent [iˈkwivələnt] adj. 等量的，相等的 n. 等价物
11. hypereutectic [ˌhaipəjuˈtektik] adj. 过共晶的
12. hypoeutectic [ˌhaipəuju (ː) ˈtektik] adj. 亚共晶的
13. outstanding [autˈstændiŋ] adj. 突出的，显著的
14. in order to ... 为了……
15. it is necessary to ... 做（干）……是必需的
16. (be) familiar with ... 熟悉，精通
17. (be) governed by ... 取决于
18. in terms of 依据，按照
19. psi = pound per square inch 每平方英寸磅

Notes

[1] In order to understand the fabricating characteristics of cast irons, it is necessary to become familiar with the characteristics of the metal and the various types and classifications that are available.

为了了解铸铁的制造特点，熟知金属的特性及各种可用的类型和分类是必要的。

句中“in order + 带 to 的不定式”表示目的状语；而在 it is necessary to become familiar with 中，it 做形式主语，后面常用动词不定式短语做真正主语；familiar with 可译为“通晓的；精通的”。

[2] Irons with a carbon equivalent of over 4.3% are called hypereutectic and are particularly good for thermal-shock resistance, such as for ingot molds.

碳当量超过 4.3% 的铸铁称为过共晶铸铁，这种铸铁的抗热冲击性能特别好，常用于铸造钢锭模。

Glossary of Terms

1. cast iron 铸铁，生铁

2. cast iron gear 铸铁齿轮
3. compacted (vermicular) graphite cast iron 蠕墨铸铁
4. white cast iron 白口铸铁
5. malleable cast iron 可锻铸铁
6. ductile (nodular, spheroidal) graphite cast iron 球墨铸铁
7. alloy cast iron 合金铸铁
8. chilled cast iron 冷硬铸铁，冷激铸铁
9. gray cast iron 灰铸铁
10. wear resisting cast iron 耐磨铸铁
11. corrosion resisting cast iron 耐蚀铸铁
12. pig iron 生铁
13. wrought iron 熟铁，锻铁
14. scrap iron 废铁（料）
15. raw material 原材料
16. ironmaking 炼铁
17. steelmaking 炼钢
18. iron ore (stone) 铁矿石
19. graphite flake 石墨片
20. carbon equivalent 碳当量
21. iron foundry 铸铁厂，铸铁车间
22. iron-carbon (Fe-Fe_3C) phase diagram 铁-碳相图
23. equilibrium diagram 平衡相图
24. eutectic composition 共晶成分
25. eutectioid temperature 共析温度
26. hypeoeutectoid steel 亚共析钢
27. hypereutectoid steel 过共析钢
28. residual stress 残余应力
29. brittle fracture 脆性断裂（断口）

Reading Materials

White Cast Irons

White cast iron is produced by cooling the cast iron rapidly so that the carbon is in the combined form as cementite. It is often referred to as chilled cast iron. The structure is very hard and brittle and the fractured surface produces a shiny appearance. If the structure contains alloying elements such as nickel, molybdenum, or chromium, martensite will be formed. Irons of this type are extremely hard and wear-resistant, and are used on hammer mills, crusher jaws, crushing rollers, balls, wear plates, etc.

Malleable Irons

Malleable iron is made from white cast iron, which has the carbon in the combined form. The white cast iron is subjected to a prolonged annealing process. It may take two days to bring the castings to a temperature of about 870℃. The castings are held at this temperature for a period of 48 to 60 hr. They are then cooled at a rate of 8 to 10 degrees per hour until the temperature is about 704℃. The temperature is held in the 677℃ to 700℃ range for a period of up to 24 hr. The furnace is then opened and allowed to cool to room temperature. During this time all the combined carbon has separated into graphite and ferrite.

A pearlitic malleable iron can be made by allowing the cooling rate to be fast enough to retain the desired amount and type of combined carbon in the matrix. In most grades, the matrix is later modified by further treatment which tempers the as-quenched structure, raises its ductility, and improves its machinability.

As the name implies, malleable irons are softer than gray irons and are therefore used where greater toughness and shock resistance are required. Malleable irons are used extensively on farm implements, automobile parts, small tools, hardware and pipe fittings.

Ductile Graphite Irons

For generations, foundrymen and metallurgists have searched for some way to transform brittle cast iron into a tough, strong material. Finally, it was discovered that magnesium could greatly change the properties of cast iron. A small amount of magnesium, about one pound per ton of iron, causes the flake graphite to take on a spheroidal shape. Graphite in a spheroidal shape presents the minimum surface for a given volume. Therefore there are less discontinuities in the surrounding metal, giving it far more strength and ductility. The processing advantages of steel, including high strength, toughness, ductility and wear resistance, can be obtained in this readily cast material.

Ductile graphite iron can be heat-treated in a manner similar to steels.

Compacted Graphite Irons

A relatively recent addition to the field of cast irons is that of compacted graphite. In 1976 a commercial treatment alloy made the process control possible. The structure and properties are generally considered to be intermediate to those of gray and ductile cast irons.

Compacted graphite iron is characterized by interconnected graphite that is intermediate to the flake graphite of gray cast iron and the nodular graphite of ductile iron. The physical properties of compacted graphite irons are to a large extent related to the interconnected graphite. While individual properties are generally intermediate to those of gray and ductile cast iron, some of the better properties of both gray and ductile are combined in compacted graphite irons.

In tensile strength, compacted graphite iron is greater than that of alloyed high-strength gray cast iron, and tensile and yield strengths approach those of ductile cast irons—depending upon the base chemistries and section size.

Unit 3 Nonferrous Metals and Alloys

Text

Nonferrous metals and alloys are playing increasingly important roles in modern *technology*. Because of their number and the fact that the properties of the individual metals vary widely, both in their relatively pure form and as base metals for alloys, they provide an almost limitless range of properties for the design engineer①. Even though they are not produced in great tonnages and are more costly than iron and steel, they make available certain important properties or combinations of properties that cannot be obtained in steels, *notably*:

(1) Resistance to corrosion.

(2) Ease of fabrication.

(3) High electrical and thermal conductivity.

(4) Light weight.

(5) Color.

While it is true that corrosion resistance can be obtained in certain ferrous alloys, several of the nonferrous alloys possess this property without requiring special and *expensive* alloying elements②. Nearly all the nonferrous alloys possess at least two of the qualities listed above, and some possess all five. For many applications, certain combinations of these properties are highly desirable, and the availability of materials that provide them directly is a strong motivation for the use of nonferrous alloys.

In most cases, the nonferrous alloys are inferior to steel in respect to strength. Also, the *modulus* of *elasticity* may be considerably lower, a fact that places them at a disadvantage where *stiffness* is a necessary property. Fabrication, however, is usually easier than steel. Those alloys with low melting points are often easy to cast, either in sand molds, *permanent* molds, or dies. Many alloys have high ductility coupled with low yield points, the ideal conditions for easy cold work and high formability. High machinability is characteristic of several nonferrous alloys. Fabrication savings can often overcome the higher cost of the nonferrous material and favor its use in place of steel. The one fabrication area in which the nonferrous alloys are somewhat inferior to steel is weldability. Due to recent developments, however, it is often possible to produce *satisfactory* weldments from the *viewpoint* of both quality and economy.

Questions

1. What are some of the important properties that may be possessed by nonferrous metals?
2. In what ways are nonferrous alloys often inferior to steel?
3. What are some of the attractive features of nonferrous metals that relate to fabrication ease?

4. Select a conductive material for the contacts of a switch or relay.

New Words and Expressions

1. technology [tekˈnɔlədʒi] n. 工艺，科技，技术
2. notably [ˈnəutəbli] adv. 显著地，特别地
3. expensive [iksˈpensiv] adj. 花费的，昂贵的
4. modulus [ˈmɔdjuləs] n. 系数，模数
5. elasticity [ilæsˈtisiti] n. 弹力，弹性
6. stiffness [ˈstifnis] n. 硬度
7. permanent [ˈpəːmənənt] adj. 永久的，持久的
8. satisfactory [ˌsætisˈfæktəri] adj. 满意的
9. viewpoint [ˈvjuː pɔint] n. 观点，观察点
10. be playing increasingly important roles 起着越来越重要的作用
11. be inferior to 低于，不及
12. in respect to 关于
13. couple with ... 与……联系在一起

Notes

[1] Because of their number and the fact that the properties of the individual metals vary widely, both in their relatively pure form and as base metals for alloys, they provide an almost limitless range of properties for the design engineer.

由于非铁金属的数量多并且各种金属的性能变化范围较宽，所以它们不论是以纯金属还是以合金的形式存在，都能为设计工程师提供几乎没有限制的性能选择范围。

句中 because of 是复合介词，也作“由于”解，但后面只能接名词短语做宾语；the fact that 引导一个同位语从句，意为“……的事实”。

[2] While it is true that corrosion resistance can be obtained in certain ferrous alloys, several of the nonferrous alloys possess this property without requiring special and expensive alloying elements.

尽管某些铁质合金的确具有耐蚀性，但一些非铁合金无需加入特殊且昂贵的合金元素即可具有良好的耐蚀性。

在分句 while it is true that corrosion resistance can be obtained in certain ferrous alloys 中，while 引导让步状语从句，表示“虽然；尽管”等含义。

Glossary of Terms

1. nonferrous metals and alloys 非铁金属及合金
2. aluminum alloys 铝合金
3. copper alloys 铜合金
4. magnesium alloys 镁合金

5. zinc alloys 锌合金
6. strength-to-weight ratio 比强度
7. nickel alloys 镍合金
8. titanium alloys 钛合金
9. low-melting alloys (lead, tin and their alloys) 低熔点合金(铅,锡及其合金)
10. high-temperature alloys 高温合金
11. heat-resistant alloys 耐热合金
12. brass (copper-zinc alloys) 黄铜
13. bronze (copper-tin alloys) 青铜
14. cupronickel (copper-nickel alloys) 白铜
15. Monel 蒙乃尔铜镍合金
16. babbitt 巴氏合金
17. corrosion resistance 耐蚀性
18. high thermal conductivity 高导热性
19. high electrical conductivity 高导电性
20. low density 低密度
21. aluminum casting alloys 铸造铝合金
22. wrought aluminum alloys 变形铝合金
23. molybdenum (Mo) 钼
24. beryuium (Be) 铍
25. tungsten (W) 钨
26. precious (noble) metal 贵金属

Reading Materials

Aluminum Alloys

Aluminum is a lightweight metal, with a density of 2.70g/cm^3 or one third the density of steel. Although aluminum alloys have relatively low tensile properties compared to steel, their strength-to-weight ratio, as defined below, is excellent.

$$\text{strength-to-weight ratio} = \text{tensile strength} / \text{density}$$

Aluminum is often used when weight is an important factor, as in aircraft and automotive applications.

Aluminum also responds readily to strengthening mechanisms. The alloys may be 30 times stronger than pure aluminum.

On the other hand, aluminum often does not display an endurance limit in fatigue, so failure eventually occurs even at rather low stresses. Because of its low melting temperature, aluminum does not perform well at elevated temperatures. Finally, aluminum alloys have a low hardness, leading to poor wear resistance.

Aluminum alloys can be subdivided into two major groups, wrought and casting alloys, based on their methods of fabrication.

Aluminum is generally very easy to machine, although the softer grades tend to form a built-up edge, resulting in poor surface finish. Thus, high cutting speeds, high rake angles and high relief angles are recommended.

Copper Alloys

There are a myriad of copper-based alloys that take advantage of all of the strengthening mechanisms that we have discussed. The copper-based alloys are heavier than iron. Although the yield strength of some alloys is high, the strength-to-weight ratio is typically less than that of aluminum or magnesium alloys. The alloys have better resistance to fatigue, creep and wear than the lightweight aluminum and magnesium alloys. Many of the alloys also have excellent ductility, corrosion resistance, and electrical and thermal conductivity.

Copper alloys are also unique in that they may be selected to produce an appropriate decorative color. Pure copper is red. However, zinc additions produce a yellow color and nickel produces a silver color.

Copper in the wrought condition can be difficult to machine because of built-up edge formation, although cast copper alloys are easy to machine. Brasses are easy to machine, especially with the addition of lead (leaded free-machining brass). Bronzes are more difficult to machine than brasses.

Magnesium Alloys

Magnesium is lighter than aluminum, with a density of $1.74 g/cm^3$, and melts at a lower temperature. Although magnesium alloys are not as strong as aluminum alloys, their strength-to-weight ratios are comparable. Consequently, magnesium alloys are used in aerospace applications, high-speed machinery, and transportation and materials handling equipment.

However, magnesium has a low modulus of elasticity and poor resistance to fatigue, creep and wear. Magnesium also poses a hazard during casting and machining, since it combines easily with oxygen and burns. Finally, the response of magnesium to strengthening mechanisms is relatively poor.

Magnesium is very easy to machine, with good surface finish and prolonged tool life. However, care should be exercised because of its high rate of oxidation and the danger of fire.

Zinc Alloys

Pure zinc is nearly as heavy as steel, melts at only 420℃, has the HCP crystal structure, and has a strength less than that of many aluminum alloys. With these properties, we might expect that zinc would seldom be used. Yet many applications are found for both wrought and cast zinc.

Because zinc recrystallizes and creeps near room temperature, it has excellent ductility. However, strain hardening is negligible. Wrought zinc is used for batteries, photoengraving plates, and roofing components, such as gutters. A special wrought alloy, Zn-22% Al, displays superplastic behavior. The unusually large deformations permit complex panels and cabinets to be formed.

Unit 4 Tool Materials

Text

The specific material selected for a particular tool is normally determined by the mechanical properties necessary for the proper operation of the tool. These materials should be selected only after a careful study and *evaluation* of the function and requirements of the proposed tool. In most applications, more than one type of material will be satisfactory, and a final choice will normally be governed by material *availability* and economic considerations.

The principal materials used for tools can be divided into three major *categories*: ferrous materials, nonferrous materials, and nonmetallic materials. Ferrous tool materials have iron as a base metal and include tool steel, alloy steel, carbon steel and cast iron. Nonferrous materials have a base metal other than iron and include aluminum, magnesium, zinc, lead, bismuth, copper and a variety of alloys. Nonmetallic materials are those materials such as woods, plastics, rubbers, epoxy resins, ceramics and diamonds that do not have a metallic base. To properly select a tool material, there are several physical and mechanical properties you should understand to determine how the materials you select will affect the function and operation of the tool[①].

Physical and mechanical properties are those characteristics of a material which control how the material will react under certain conditions. Physical properties are those properties which are natural in the material and cannot be permanently altered without changing the material itself. These properties include weight, color, *thermal* and electrical conductivity, rate of thermal expansion and melting point. The mechanical properties of a material are those properties which can be permanently altered by thermal or mechanical treatment. These properties include strength, hardness, wear resistance, toughness, brittleness, plasticity, ductility, malleability and modulus of elasticity.

From a use *standpoint*, tool steels are utilized in working and shaping basic materials such as metals, plastics and woods into desired forms. From a composition standpoint, tool steels are carbon alloy steels which are capable of being hardened and tempered. Some desirable properties of tool steels are high wear resistance and hardness, good heat resistance, and *sufficient* strength to work the materials. In some cases, *dimensional stability* may be very important. Tool steels also must be economical to use and be capable of being formed or machined into the desired shape for the tool.

Since the property requirements are so special, tool steels are usually melted in electric furnaces using careful *metallurgical* quality control. A great effort is made to keep *porosity*, *segregation*, *impurities* and nonmetallic inclusions to as low a level as possible[②]. Tool steels are subjected to careful *macroscopic* and *microscopic inspections* to ensure that they meet strict "tool steel" specifications.

Although tool steels are a relatively small percentage of total steel production, they have a strategic position in that they are used in the production of other steel products and engineering materials. Some applications of tool steels include drills, deep drawing dies, shear *blades*, punches, extrusion dies and cutting tools.

For some applications, especially where extremely high-speed cutting is important, other tool materials such as sintered carbide products are a more economical alternative to tool steels. The exceptional tool performance of sintered carbides results from their very high hardness and high compressive strength. Other tool materials are being used more and more often industrially.

Questions

1. What is meant by the term "physical properties of a material"?
2. What are the mechanical properties of a material?
3. What makes a material either ferrous or nonferrous?
4. What are some applications of tool steels?

New Words and Expressions

1. evaluation [iˌvæljuˈeiʃn] n. 评价，估价，鉴定
2. availability [əˌveiləˈbiləti] n. 存在，有效性，利用率
3. category [ˈkætəgəri] n. 种类，范畴，类型
4. thermal [ˈθəːml] adj. 热（量）的，由热造成的 n. 上升暖气流
5. standpoint [ˈstændpɔint] n. 观点，立场
6. sufficient [səˈfiʃnt] adj. 充分的，足够的
7. dimensional [daiˈmænʃənl] adj. 尺寸的，量纲的
8. stability [stəˈbiliti] n. 稳定（性），安定度
9. metallurgical [ˌmetəˈləːdʒikəl] adj. 冶金（学）的
10. porosity [pɔːˈrɔsiti] n. 多孔（性），孔隙度，疏松（度）
11. segregation [ˌsegriˈgeiʃən] n. 分离，偏析
12. impurity [imˈpjuəriti] n. 杂质
13. macroscopic [ˌmækrəuˈskɔpik] adj. 宏观的，肉眼可见的
14. microscopic [ˌmaikrəˈskɔpik] adj. 显微镜的，微观的
15. inspection [inˈspekʃən] n. 检查，调查，视察
16. blade [bleid] n. 刀口，刀片，刀身
17. under certain conditions 在一定条件下

Notes

[1] To properly select a tool material, there are several physical and mechanical properties you should understand to determine how the materials you select will affect the function and operation of the tool.

为了合理选择工具材料，你应当掌握材料的一些物理性能和力学性能，以便确定所选材

料对工具的功能和操作会有何影响。

不定式 to properly select a tool material 放在句首用作目的状语；you select 作为定语从句修饰前面的 the materials。

[2] A great effort is made to keep porosity, segregation, impurities and nonmetallic inclusions to as low a level as possible.

(对于工具钢的冶炼,) 要最大限度地降低钢中的气孔、偏析、杂质以及非金属夹杂物的含量。

这里的 a great effort is made 是 make a great effort 的被动形式，意为“尽一切力量”。

Glossary of Terms

1. carbide tool 硬质合金刀具
2. carbon steel 碳素钢
3. alloy steel 合金钢
4. carbon tool steel 碳素工具钢
5. alloy tool steel 合金工具钢
6. high-carbon steel 高碳钢
7. medium-carbon steel 中碳钢
8. low-carbon steel 低碳钢
9. high-alloy steel 高合金钢
10. low-alloy steel 低合金钢
11. alloy cast iron 合金铸铁
12. stainless steel 不锈钢
13. cast steel 铸钢
14. die block steel 模具钢
15. die material 模具材料
16. free cutting (machining) steel 易切削钢
17. shock resistant tool steel 抗冲击工具钢
18. cold work tool (die) steel 冷作工具（模具）钢
19. hot work tool (die) steel 热作工具（模具）钢
20. special purpose steel 特殊用途钢
21. killed steel 镇静钢
22. semi-killed steel 半镇静钢
23. high speed steel 高速钢
24. uncoated carbides 无涂层硬质合金
25. coated carbides 涂层硬质合金
26. polycrystalline cubic boron nitride 多晶立方氮化硼
27. diamond 金刚石
28. high-strength low-alloy steels (HSLA) 高强度低合金钢

29. tungsten carbide 碳化钨，硬质合金
30. cemented (sintered) carbides 渗碳（烧结）硬质合金

Reading Materials

Carbon Steels

Carbon steels are used extensively in tool construction. Carbon steels are those steels which only contain iron and carbon, and small amounts of other alloying elements. Carbon steels are the most common and least expensive type of steel used for tools. The three principal types of carbon steels used for tooling are low-carbon, medium-carbon and high-carbon steels. Low-carbon steel contains carbon between 0.05% and 0.3%. Medium-carbon steel contains carbon between 0.3% and 0.7%. And high-carbon steel contains carbon between 0.7% and 1.5%. As the carbon content is increased in carbon steel, the strength, toughness and hardness are also increased when the metal is heat treated.

Low-carbon steels are soft, tough steels that are easily machined and welded. Due to their low carbon content, these steels cannot be hardened except by case hardening. Low-carbon steels are well suited for the following applications: tool bodies, handles, die shoes, and similar situations where strength and wear resistance are not required.

Medium-carbon steels are used where greater strength and toughness is required. Since medium-carbon steels have a higher carbon content, they can be heat treated to make parts such as studs, pins, axles and nuts. Steels in this group are more expensive as well as more difficult to machine and weld than low-carbon steels.

High-carbon steels are the most hardenable type of carbon steel and are used frequently for parts where wear resistance is an important factor. Other applications where high-carbon steels are well suited include drill bushings, locators and wear pads. Since the carbon content of these steels is so high, parts made from high-carbon steel are normally difficult to machine and weld.

Carbon steels have a wide range of machinability, depending on their ductility and hardness. If too ductile, chip formation can produce built-up edge, leading to poor surface finish; if the steel is too hard, it can cause abrasive wear of the tool because of the presence of carbides in the steel. Cold-worked carbon steels are desirable from a machinability standpoint.

Alloy Steels

Alloy steels are basically carbon steels with additional elements added to alter the characteristics and bring about a predictable change in the mechanical properties of the alloyed metal. Alloy steels are not normally used for most tools due to their increased cost, but some have been found favor for special applications. The alloying elements used most often in steels are manganese, nickel, molybdenum and chromium.

Another type of alloy steel frequently used for tooling applications is stainless steel. Stainless steel is a term used to describe high chromium and nickel-chromium steels. These steels are used for tools which must resist high temperatures and corrosive atmospheres. Some high chromium steels can be hardened by heat treatment and are used where resistance to wear, abrasion and corrosion are required. Typical applications where a hardenable stainless steel is sometimes preferred are plastic injection molds. Here the high chromium content allows the steel to be highly polished and prevents deterioration of the cavity from heat and corrosion.

Alloy steels can have a wide variety of compositions and hardnesses. Consequently, their machinability cannot be generalized, although they have higher levels of hardness and other mechanical properties. An important trend in machining these steels is hard turning. Using polycrystalline cubic-boron-nitride cutting tools, alloy steels at hardness levels of 45 to 65 HRC can be machined with good surface finish, integrity and dimensional accuracy.

Cutting Tool Materials

In the cutting process, the cutting tool takes part in the cutting work directly and produces the machined surface, so that tool materials have an important effect on the productivity, cost and quality of metal cutting. Therefore, we should choose and use tool materials rightly.

Performances for cutting tool materials. In metal cutting, the cutting parts of the tool are subjected to force, temperature and friction, thereby it should meet the following basic requirements: ①High hardness and wear resistance; ②Sufficient strength and toughness; ③High heat resistance; ④Good technological properties; ⑤Good economy conditions.

Frequently-used cutting tool materials. At present, the most commonly used cutting materials are high-speed steels and carbide alloys. Carbon tool steels such as T10A, T12A, alloy tool steels such as 9SiCr, CrWMn, are only used in some hand tools or cutting tools working at low cutting speed, due to their poor heat resistance.

(1) High speed steels (HSS). There are high alloy tool steels having more tungsten, molybdenum, chromium, vanadium and other elements so that they possess higher heat resistance and strength, toughness as well as wear resistance, can work from 500℃ to 650℃. Due to their good technological properties, they are the main materials of complex tools, such as drills, formed cutters, broaches, gear cutting tools, etc.

According to applications, high-speed steels can be divided into plain high-speed steels (tungsten steel W18Cr4V and tungsten molybdenum steel W6Mo5Cr4V2) and high property high-speed steels (high carbon high-speed steel 9W18Cr4V, high vanadium high-speed steel W6MoCr4V3, cobalt high-speed steel W6MoCr4V2Co8, super hard high-speed steel W2Mo9Cr4VCo8, etc.)

(2) Carbide alloys. The carbide alloys are made of metallic carbide powder (WC, TiC, etc.) and bonding mediums (Co, Mo, Ni) through powder metallurgy. It is usually referred to as sintered carbide and sometimes it is called cemented carbide.

Other tool materials. (1) Coated tools. The carbide alloys or HSS on which a layer of substance with high melting point, high hardness and wear resistance, such as TiC, TiN, Al_2O_3, etc., is coated, are called coated carbide alloys. It can be used in finishing and semi-finishing all work materials except titanium alloys and austenitic stainless steel. (2) Ceramics. There are Al_2O_3 and Al_2O_3 TiC ceramic tools, both of which are made of their powders through sintering to produce fine grain structure and maximum density. Ceramic tools possess very high hardness (91 ~ 95HRA), chemical stability, anti-adhesiveness, heat resistance and wear resistance, and have less affinities for the workpiece. They can withstand cutting temperatures as high as 1200° C, may be used to machine steel or cast iron and are suited to turning and milling processes. (3) Diamond. Diamond is the hardest material. It may be divided into natural and artificial diamonds. A diamond tool is qualified to machine not only high hardness and wear resistant materials such as carbide alloy, ceramics, etc. , but also non-ferrous metals. (4) Cubic boron nitride (CBN). It is formed from the transformation of hexagonal boron nitride due to catalysis under high temperature and high pressure. Its hardness (8000 ~ 9000HV) is only inferior to diamond. Its heat stability is much better than diamond so that it can withstand cutting temperature in the range of 1300 ~ 1500°C. It may be used to machine high temperature alloys, hardness steels, chill cast iron and other materials, and is suitable for turning, milling, boring and reaming, and is also used as grinding tools.

Unit 5 Heat Treatment of Tool Steels

Text

The purpose of heat treatment is to control the properties of a metal or alloy through the alteration of the structure of the metal or alloy by heating it to definite temperatures and cooling at various rates. This combination of heating and controlled cooling determines not only the nature and *distribution* of the microconstituents, which in turn determine the properties, but also the *grain* size①.

Heat treatment should improve the alloy or metal for the service intended. Some of the various purposes of heat treatment are as follows:

(1) To remove *strains* after cold working.

(2) To remove internal stresses such as those produced by drawing, bending, or welding.

(3) To increase the hardness of the material.

(4) To improve machinability.

(5) To improve the cutting capabilities of tools.

(6) To improve wear-resisting properties.

(7) To soften the material, as in *annealing*.

(8) To improve or change properties of a material, such as *corrosion* resistance, heat resistance, *magnetic* properties, or others as required.

Treatment of ferrous materials. Iron is the major *constituent* in the steels used in tooling, to which carbon is added in order that the steel may harden. Alloys are put into steel to enable it to develop properties not possessed by plain carbon steel, such as the ability to harden in oil or air, increased wear resistance, higher toughness, and greater safety in hardening.

Heat treatment of ferrous materials involves several important operations which are *customarily* referred to under various headings, such as normalizing, *spheroidizing* annealing, stress relieving, annealing, hardening capacity, tempering and surface hardening.

Normalizing. It involves heating the material to a temperature of about 100 ~ 200 ℉ (55 ~ 100℃) above the critical range and cooling in still air. This is about 100 ℉ (55℃) over the regular hardening temperature.

The purpose of normalizing is usually to refine grain structures that have been *coarsened* in forging. With most of the medium-carbon forging steels, alloyed and unalloyed, normalizing is highly recommended after forging and before machining to produce more *homogeneous* structures, and in most cases, improved machinability②.

High-alloy air-hardened steels are never normalized, since to do so would cause them to harden and defeat the primary purpose.

Spheroidizing annealing. It is a form of annealing which in the process of heating and cooling steel, produces a rounded or *globular* form of carbide—the hard constituent in steel.

Tool steels are normally spheroidized to improve machinability. It is accomplished by heating to a temperature to 1380 ~1400 ℉ (749 ~760℃) for carbon steels and higher for many alloy tool steels, holding at heat one to four hours, and cooling slowly in the furnace.

Stress relieving. This is a method of relieving the internal stresses set up in steel during forming, cold working, and cooling after welding or machining. It is the simplest heat treatment and is accomplished merely by heating to 1200 ~ 1350 ℉ (649 ~732℃) followed by air or furnace cooling.

Large dies are usually roughed out, then stress-relieved and finish-machined. This will minimize the change of shape not only during machining but during subsequent heat treatment as well. Welded sections will also have locked-in stresses owing to a combination of differential heating and cooling cycles as well as to changes in cross section. Such stresses will cause considerable movement in machining operations.

Annealing. The process of annealing consists of heating the steel to an *elevated* temperature for a definite period of time and, usually, cooling it slowly. Annealing is done to produce homogenization and to establish normal equilibrium conditions, with corresponding characteristic properties.

Tool steel is generally purchased in the annealed condition. Sometimes it is necessary to rework a tool that has been hardened, and the tool must then be annealed. For this type of anneal, the steel is heated slightly above its critical range and then cooled very slowly.

Hardening. This is the process of heating to a temperature above the critical range, and cooling rapidly enough through the critical range to *appreciably* harden the steel.

Tempering. This is the process of heating quenched and hardened steels and alloys to some temperature below the lower critical temperature to reduce internal stresses set up in hardening.

Surface hardening. The addition of carbon to the surfaces of steel parts and the subsequent hardening operations are important phases in heat treatment. The process may involve the use of molten sodium *cyanide* mixtures, pack carburizing with activated solid material such as charcoal or coke, gas or oil *carburizing*, and dry cyaniding.

Questions

1. What process is used to remove the internal stresses created during a hardening operation?
2. What heat treatment process makes the metallic carbides in a metal form into small rounded globules?
3. What are the main purposes of heat treatment?
4. How many heat treatment processes are involved in ferrous materials?

New Words and Expressions

1. distribution [ˌdistriˈbjuːʃən] n. 分配，分布
2. grain [grein] n. 晶粒，粒度 vt. 使成细粒 vi. 形成粒状

3. strain [strein] vt. 应变，张力，变形 n. 压力，张力，应力
4. annealing [əˈniːliŋ] n. 退火，韧化，缓冷
5. corrosion [kəˈrəuʒən] n. 腐蚀，锈，铁锈
6. magnetic [mægˈnetik] adj. 磁的，有吸引力的
7. constituent [kənˈstitjuənt] n. 成分，要素 adj. 组成的
8. customarily [ˈkʌstəmərili] adv. 通常，习惯上
9. spheroidizing [ˈsfiərɔidaiziŋ] n. 球化处理
10. coarsen [ˈkɔːsn] vt. 使粗，粗化 vi. 变粗
11. homogeneous [ˌhɔməˈdʒiːnjəs] adj. 同种的，同性的，均匀的
12. globular [ˈglɔbjulə] adj. 球状的
13. elevate [ˈeliveit] vt. 抬起，举起，使升高
14. appreciable [əˈpriːʃiəbl] adj. 可估计的，明显的
15. cyanide [ˈsaiənaid] n. 氰化物
16. carburize [ˈkɑːbjuraiz] vt. 渗碳
17. grain size 晶粒尺寸
18. internal stress 内应力
19. corrosion resistance 耐腐蚀
20. heat resistance 耐热
21. magnetic property 磁性能
22. (be) referred to 把……归因于，参考，认为……由于
23. critical range 临界范围

Notes

[1] This combination of heating and controlled cooling determines not only the nature and distribution of the microconstituents, which in turn determine the properties, but also the grain size.

(在热处理过程中,) 加热与控制冷却相结合的方法不仅决定着材料中影响材料性能的金相显微组织的性质和分布，而且也决定了材料内部晶粒的大小。

句中第一个 and 连接的是 heating 和 controlled cooling，第二个 and 连接的是 nature 和 distribution，而 which in turn determine the properties 为非限定性定语从句，修饰 the nature and distribution。

[2] With most of the medium-carbon forging steels, alloyed and unalloyed, normalizing is highly recommended after forging and before machining to produce more homogeneous structures, and in most cases, improved machinability.

对于大多数中碳锻钢，不管它是否合金化，在锻造后及加工前通常十分推荐进行正火处理，以获得更均匀的组织，并且在大多数情况下材料的切削加工性能也会得到改善。

句首的介词 with 短语意为“对……来说”；in most cases 可译为“在大多数情况下”。

Glossary of Terms

1. hardenability 淬透性

2. hardening capacity 淬硬性（硬化能力）
3. case hardening 表面硬化，渗碳
4. hardness profile 硬度分布（硬度梯度）
5. heat treatment procedure 热处理规范
6. heat treatment installation (equipment) 热处理设备
7. heat treatment furnace 热处理炉
8. heat treatment cycle 热处理工艺周期
9. heat time 加热时间
10. normalizing 正火
11. tempering 回火
12. quenching 淬火
13. austenite 奥氏体
14. bainite 贝氏体
15. martensite 马氏体
16. heating curve 加热曲线
17. high temperature carburizing 高温渗碳
18. high temperature tempering 高温回火
19. isothermal transformation (IT) 等温转变
20. isothermal annealing 等温退火
21. interrupted ageing treatment 分级时效处理
22. local heat treatment 局部热处理
23. overheated structure 过热组织
24. pack carburizing 固体渗碳
25. oxynitrocarburizing 氧氮碳共渗
26. partial annealing 不完全退火
27. spheroidized structure 球化组织
28. recrystallization temperature 再结晶温度
29. stress relieving 去应力
30. age hardening 时效硬化
31. time-temperature-transformation (TTT) diagram 时间温度转变图，也叫 C 曲线
32. surface hardening 表面淬火
33. surface hardness 表面硬度
34. solid solution strengthening 固溶强化
35. continuous cooling transformation (CCT) curve 连续冷却转变曲线
36. cooling curve 冷却曲线

Reading Materials

Heat Treatment of Die Steels

Although alloy steels contain elements such as chromium, molybdenum and vanadium, two constituents are essential for heat treatment: iron, termed ferrite in metallography, and carbon, which combines with iron to form cementite, the hard intermetallic compound Fe_3C. These two Constituents form a eutectoid structure known as pearlite when the steel is cooled slowly enough to reach equilibrium, but by rapid cooling the steel is hardened. When such a quenched steel is tempered, structures with mechanical properties intermediate between those of the slowly cooled and the quenched conditions are formed.

In recent years there has been a greater understanding of the complex structural changes taking place during heat treatment, with the help of phase transformation diagrams. Use of these diagrams can lead to better control of the heat treatment cycle which in turn will ensure that optimum properties and maximum die life are achieved.

Surface Treatments for Steels

During the past 20 years, several processes have been introduced to obtain enhanced surface hardness of steels. Some of them have developed from case carburizing and nitriding, to obtain shorter processing times with better environmental control and improved properties. Various salt bath processes have been used and now a wide range of new methods is available.

In the die casting industry surface treatments are applied to steels to improve the properties of nozzles, ejector pins, cores and shot sleeves, to provide maximum resistance to erosion, pitting and soldering. Treatment of die cavities has received only limited acclaim, because the complex thermal patterns produced on large die components lead to stresses which are sufficiently high to break through the thin surface treated layers, leading to premature failure. Experience in drop forging has also indicated that surface treatments of their dies have not been particularly successful.

Thermochemical treatments are applied to casting die components; the surface chemistry of the steel is modified by the introduction of nitrogen, carbon, and sometimes other elements; the processes are of the main types listed below.

(1) Nitriding.

(2) Nitrocarburizing such as Tufftride, Sulfinuz and Suraulf.

(3) Metallizing such as boronizing and the Toyota diffusion process.

(4) Carburizing and carbonitriding.

Heat Treatments to Increase Strength

Six major mechanisms are available to increase the strength of metals:

① Solid solution hardening. ②Strain hardening. ③Grain-size refinement. ④Precipitation hardening. ⑤Dispersion hardening. ⑥Phase transformations. All can be induced or altered by heat treatment but not all can be applied to any given metal.

In solid solution hardening, a base metal dissolves other atoms in solid solution, either as substitutional solutions, where the new atoms occupy sites on the regular crystal lattice, or as interstitial solutions, where the new atoms squeeze into 'holes' in the base lattice. The amount of strengthening depends on the amount of dissolved solute and the size difference of the atoms involved. Distortion of the host structure makes dislocation motion more difficult.

Strain hardening produces increased strength by plastic deformation under cold-working conditions.

Because grain boundaries act as barriers to dislocation motion, a metal with small grains will tend to be stronger than the same metal with larger grains. Thus grain-size refinement can be used to increase strength, except at elevated temperatures where failure is by a grain-boundary diffusion-controlled creep mechanism. Grain-size refinement is one of the few processes capable of improving both strength and ductility.

Phase transformation strengthening involves alloys which can be heated to form a single high-temperature phase and subsequently transformed to one or more low-temperature phases upon cooling. Where phase transformation is used to increase strength, the cooling is usually rapid and the phases produced are nonequilibrium in nature.

Chapter 3 Engineering Materials Forming Methods

Unit 1 Foundry Processes

Text

Several different methods, such as *casting*, *molding*, forming, powder metallurgy and machining, are available to shape metals into useful products. One of the oldest processes is casting.

Founding, or casting, is the process of forming objects by *pouring* liquid or *viscous* material into a prepared mold or form. A casting is an object formed by allowing the material to solidify. A foundry is a collection of the necessary materials and the equipments to produce a casting. Practically all metal is initially cast. The ingot from which a wrought metal is produced is the first cast in an ingot mold. A mold is the container that has the *cavity* (or cavities) of the shape to be cast. Liquids may be poured; some liquids and all viscous plastic materials are forced under pressure into the molds.

In casting, a solid is melted, heated to a proper temperature, and treated to produce a desired chemical *composition*. The molten material, generally metal, is then poured into a cavity or a mold which contains it in proper shape during *solidification*. Thus, in a single step, simple or complex shapes can be made from any metal that can be melted, with the resulting product having *virtually* any *configuration* the designer desires for best resistance to working stresses, minimal directional properties, and usually, a pleasing appearance①.

Founding is one of the oldest industries in the metalworking field and dates back to approximately 4, 000B. C②. Since this early age, many methods have been employed to cast various materials. In this chapter, sand casting and its *ramifications* receive the first attention because they are the most used methods; over 90% of all castings are sand castings. Sand casting is best suited for iron and steel at their high melting temperatures but also predominates for aluminum, brass, bronze and magnesium. Other processes of commercial importance are treated, in most cases those for *nonferrous* metals using *permanent* molds.

The elements necessary for the production of sound castings are considered throughout this chapter. These include molding materials, molding equipment, tools, *patterns*, melting equipment, etc. These basic *ingredients* must be combined in an orderly sequence for the production of a sound casting[③].

Castings have specific important engineering properties; these may be *metallurgical*, physical, or economic. Castings are often cheaper than *forgings* or *weldments*, depending on the quantity, type of material, and cost of patterns as compared with the cost of *dies* for forgings and the cost of *jigs* and *fixtures* for weldments. Where this is the case, they are the logical choices for engineering structures and parts.

Questions

1. What are the advantages of casting as compared with other processes?
2. What is the casting? Why is casting an important manufacturing process?
3. Describe the foundry processes.
4. Discuss the elements necessary for the production of sound castings.
5. Which casting method is the most commonly used?

New Words and Expressions

1. casting [ˈkɑːstiŋ] n. 铸件，铸造
2. mold [məuld] n. 型，模子，铸型 vt. 浇铸，塑造
3. pour [pɔː] v. & n. 浇注，灌注
4. viscous [ˈviskəs] adj. 黏性的，胶粘的
5. cavity [ˈkæviti] n. 型腔，空腔
6. composition [kɔmpəˈziʃən] n. 组成，成分
7. solidification [ˌsɔlidifiˈkeiʃən] n. 凝固
8. virtual [ˈvəːtjuəl] adj. 实际上的，有效的
9. configuration [kənˌfigjuˈreiʃən] n. 外形，形状，配置
10. ramification [ˌræmifiˈkeiʃən] n. 分枝，衍生物
11. nonferrous [ˈnɔnˈferəs] adj. 不含铁的，非铁的
12. permanent [ˈpəːmənənt] adj. 永久的，持久的
13. pattern [ˈpætən] n. 模范，模型
14. ingredient [inˈgriːdiənt] n. 成分，因素
15. metallurgical [ˌmetəˈləːdʒikəl] adj. 冶金学的
16. forging [ˈfɔːdʒiŋ] n. 锻造，锻件
17. weldment [ˈweldmənt] n. 焊（接）件
18. die [dai] n. 模，冲模
19. jig [dʒig] n. 夹具
20. fixture [ˈfikstʃə] n. 夹紧装置，夹具
21. pour into 将……浇入……
22. chemical composition 化学成分
23. in the field of 在……方面，在……范围内

Notes

[1] Thus, in a single step, simple or complex shapes can be made from any metal that can be melted, with the resulting product having virtually any configuration the designer desires for best resistance to working stresses, minimal directional properties, and usually, a pleasing appearance.

因此，不论铸件的形状是简单还是复杂，可由任何能够被熔化的金属通过上述步骤来实现铸造。且最终铸件可实现设计师所期望的形状，并具有最佳的抗加工应力和最小的定向性，往往还具有令人满意的外观。

句中 that can be melted 是定语从句，修饰 any metal，可译为：“凡是能被熔化的金属”；句中 with the resulting product having ... 为“with + 名词 + 分词”构成的独立主格结构，表示陪衬性的动作或附加说明；be made from 可译为“由……制成”。

[2] Founding is one of the oldest industries in the metalworking field and dates back to approximately 4000B. C.

在金属加工领域中，铸造是最古老的行业之一，可追溯到大约公元前4000 年。

句中的 date back to 指“从……时就有，回溯到，远在……（年代）”。

[3] These basic ingredients must be combined in an orderly sequence for the production of a sound casting.

为了生产出合格的铸件，这些基本要素必须按一定顺序结合起来。

Glossary of Terms

1. casting, founding, foundry 铸造
2. sand casting 砂型铸造
3. special casting 特种铸造
4. caster, founder, foundry worker 铸工
5. foundryman 铸造工作者
6. foundry shop 铸造车间
7. foundry technology, casting process 铸造工艺
8. foundry materials 铸造用材料
9. foundry equipment, foundry facilities 铸造设备
10. casting alloys 铸造合金
11. shell-mold casting 壳型铸造
12. full-mold casting (expendable pattern casting) 实型铸造（消失模铸造）
13. plaster-mold casting 石膏型铸造
14. investment casting, lost wax casting 熔模铸造（失蜡铸造）
15. permanent-mold casting, metal mold casting 金属型铸造
16. die casting, pressure die casting 压力铸造（压铸）
17. hot chamber die casting machine 热室压铸机
18. cold chamber die casting machine 冷室压铸机

19. continuous casting 连续铸造
20. precision casting 精密铸造
21. ceramic mold casting 陶瓷型铸造
22. low-pressure casting 低压铸造
23. centrifugal casting 离心铸造
24. vacuum casting 真空铸造
25. squeeze casting 挤压铸造
26. slush casting 空心铸件
27. semisolid-metal forming 半固态金属成形
28. dimensional tolerance of casting 铸件尺寸公差
29. machining allowance of casting 铸件机械加工余量
30. casting defect 铸件缺陷

Reading Materials

Design Considerations

As in all manufacturing operations, certain guidelines and design principles pertaining to casting have been developed over many years. These principles were established primarily through practical experience, but new analytical methods, process modeling, and computer-aided design and manufacturing techniques are now coming into wider use, improving productivity and the quality of castings and resulting in significant cost savings.

The guidelines that follow apply to all types of casting generally. The most significant design considerations are identified and addressed.

1. Corners, angles and section thickness. Sharp corners, angles, and fillets should be avoided, because they act as stress raisers and may cause cracking and tearing of the metal (as well as of the dies) during solidification. Fillet radii should be selected to reduce stress concentrations and to ensure proper liquid-metal flow during the pouring process. On the other hand, if the fillet radii are too large, the volume of the material in those regions is also large and, consequently, the rate of cooling is lower.

Section changes in castings should be smoothly blended into each other. The location of the largest circle that can be inscribed in a particular region is critical so far as shrinkage cavities are concerned. Because the cooling rate in regions with larger circles is lower, they are called hot spots. These regions could develop shrinkage cavities and porosity.

2. Flat areas. Large flat areas (plain surfaces) should be avoided, because they may warp during cooling because of temperature gradients or develop poor surface finish (because of uneven flow of metal during pouring). Flat surfaces can be broken up with ribs and serrations.

3. Shrinkage. To avoid cracking of the casting, there should be allowances for shrinkage during solidification. In castings with intersecting ribs, the tensile stresses can be reduced by staggering the ribs or by changing the intersection geometry.

Pattern dimensions should also provide allowances for shrinkage of the metal during solidification and cooling. Allowances for shrinkage is also known as patternmaker's shrinkage allowances.

4. Parting line. The parting line is the line, or plane, separating the upper (cope) and lower (drag) halves of molds. In general, it is desirable for the parting line to be along a flat plane, rather than contoured. Whenever possible, the parting line should be at the corners or edges of castings, rather than on flat surfaces in the middle of the casting, so that the flash at the parting line (material squeezing out between the two halves of the mold) will not be as visible.

The location of the parting line is important because it influences mold design, ease of molding, number and shape of cores, method of support, and the gating system.

5. Draft. A small draft (taper) is provided in sand-mold patterns to enable removal of the pattern without damaging the mold. Typical draft range from 5mm/m to 15mm/m. Depending on the quality of the pattern, draft angles usually range from 0.5° to 2°.

6. Dimensional tolerances. Dimensional tolerances depend on the particular casting process, size of the casting, and type of pattern used. Tolerances are smallest within one region of the mold, but because they are cumulative, increase between different regions of the mold. Tolerances should be as wide as possible, within the limits of good part performance; otherwise, the cost of the casting increases. In commercial practice, tolerances are usually in the range of ±0.8mm for small castings and increase with the size of castings. Tolerances for large castings, for instance, may be ±6mm.

7. Machining allowances. Because most expendable-mold castings require some additional finishing operations, such as machining, allowances should be made in casting design for these operations. Machining allowances, which are included in pattern dimensions, depend on the type of casting and increase with the size and section thickness of castings. Allowances usually range from about 2mm to 5mm for small castings to more than 25mm for large castings.

8. Residual stresses. The different cooling rates within the body of a casting cause residual stresses. Stress relieving may thus be necessary to avoid distortions in critical applications.

Sand Casting

In sand casting, sand is the material used for making both molds and cores. A pattern, commonly of wood or metal, is needed to form the cavity into which molten metal will be poured. Sand molds serve for one pouring only and are necessarily broken up when the castings are removed.

Good castings cannot be produced without good molds. Because of the importance of the mold, casting processes are often described by the material and method employed for the mold. Thus sand casting may be made in ①green sand molds, ②dry sand molds, ③core sand molds, ④loam molds, ⑤shell molds, and ⑥cement-bonded molds. The major methods of making these molds are called ①bench molding, ②machine molding, ③floor molding, and ④pit molding.

Figure 3-1-1 shows two-part mold ready for pouring. The term mold is used to include the entire assembly as shown. Molten metal is poured into the pouring basin. It flows downward through the gate sprue into a runner or other enlargement at the parting surface and then enters the mold cavity through the gate or gates. The principal reason for having a riser is to provide molten metal to compensate for metal shrinkage in the casting during freezing. The gating and risers are removed from the casting in the cleaning department and later remelted.

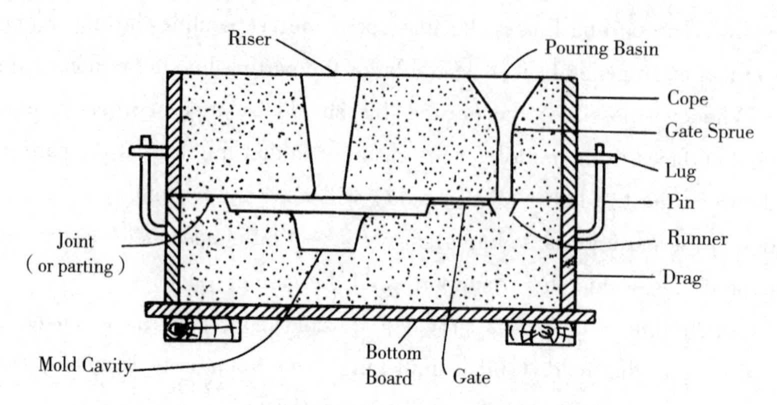

Figure 3-1-1 Cross-sectional view through a typical sand mold

Advantages:

(1) Sand casting provides a great deal of flexibility. There are very few limitations on casting size or shape.

(2) Molten material flows into any small section in the mold cavity and as such any intricate shapes internal or external can be made with the casting process.

(3) Fast production rates can be achieved for small castings. Sand casting provides one for the quickest ways to get from the drawing board to mass production.

(4) It is possible to cast practically any material ferrous or nonferrous.

Limitations:

(1) Sand casting is not used where the highest dimensional accuracy is desired. Surfaces usually need finishing operations.

(2) Sand casting process is labour intensive to some extent and therefore many improvements are aimed at it such as machine molding and foundry mechanization.

(3) With some materials it is often difficult to remove defects arising out of the moisture present in sand casting.

Unit 2 Soldering and Welding

Text

There are a number of methods of *joining* metal *articles* together, depending on the type of metal and the strength of the *joint* which is required①.

Soldering is the process of joining metals by a third metal to be applied in the molten state. Solder consists of tin and lead, while bismuth and cadmium are often included to lower the melting point. One of the most important operations in soldering is that of cleaning the surfaces to be joined, this may be done by some acid cleaner②. Although the oxides are removed by the cleaning operation, a new oxide coating forms immediately after cleaning, thus preventing the solder to unite with the surface of the metal. *Flux* is used to remove and prevent oxidation of the metal surfaces to be soldered, allowing the solder to flow freely and unite with the metal③. Zinc chloride is the best flux to use for soldering most ferrous and nonferrous metals. For soldering aluminum, stearine or vaseline are to be used as fluxes.

Soldering gives a satisfactory joint for light articles of steel, copper or brass, but the strength of a soldered joint is rather less than a joint which is brazed, *riveted* or welded. These methods of joining metal are normally *adopted* for strong permanent joint.

In some cases it may be necessary to connect metal surfaces by means of a hard solder which fuses at a high temperature. This kind of soldering is called brazing.

Welding is a metallurgical *fusion* process. Here, the interfaces of the two parts to be joined are brought to a temperature above the melting point and then allowed to solidify so that a permanent joining takes place. Because of the permanent nature of the joint and the strength being equal to or sometimes greater than that of the parent metal, welding becomes one of the most extensively used fabrication method. Welding is not only used for making structures but also for repair work such as the joining of broken castings. Many parts of machines, automobiles, airplanes, ships, bridges and buildings are welded.

Oxyacetylene welding is the heating of two pieces of metal with a *flame* which burns a mixture of oxygen and acetylene gas. The oxygen and acetylene gas are kept in two separate steel tanks from which they flow to a *torch*, there the two gases mix and then pass into the flame. It is the hottest flame known for ordinary use, its temperature is about 3480℃. The oxyacetylene flame may be also used to cut iron and steel.

Electric *arc* welding is the heating of two pieces of metal to be welded by electricity. This is the hottest arc that can be obtained for engineering purposes, it is about 4000℃. *Spot* welding is welding two pieces of metal in spots with electricity and is done with machine called spot welder.

A forged weld is made by softening the ends of two metal pieces in a furnace and then hammering them together.

Many different energy sources can be used for welding, including a gas flame, an electric arc, a laser, an electron beam, friction and ultrasound. While often an industrial process, welding can be done in many different environments, including open air, underwater and in space. Regardless of location, however, welding remains dangerous, and precautions must be taken to avoid burns, electric shock, poisonous fumes and overexposure to ultraviolet light.

Questions

1. What is soldering, brazing and welding?
2. How does soldering differ from welding?
3. How does soldering differ from brazing?
4. How many welding methods do you know so far?
5. Before welding, why must the surface of metals be cleaned first?
6. Describe oxyacetylene welding.

New Words and Expressions

1. soldering [ˈsɔldəriŋ] n. 软钎焊；低温焊接
2. welding [ˈweldiŋ] n. 焊接法，粘结
3. join [dʒɔin] vt. 连接，结合 n. 连接
4. article [ˈɑːtikl] n. 文章，物品
5. joint [dʒɔint] n. 接头；连接方法；接缝
6. flux [flʌks] vi. 熔化 vt. 使熔融 n. 钎剂
7. rivet [ˈrivit] n. 铆钉 v. 固定
8. adopt [əˈdɔpt] vt. 采用，采纳
9. fusion [ˈfjuːʒən] n. 熔化，熔合，熔接
10. oxyacetylene [ˌɔksiəˈsetəliːn] n. 氧乙炔（气）
11. flame [fleim] n. 火焰
12. torch [tɔːtʃ] n. 火炬，焊炬 vt. 用火炬点燃
13. arc [ɑːk] n. 弧，弧光；电弧
14. spot [spɔt] n. 斑点，地点；点焊
15. electrode [iˈlektrəud] n. 电极；焊条
16. pressure welding 压力焊接，压焊
17. be attached to 依附上，贴上
18. to a great degree 很大程度上

Notes

[1] There are a number of methods of joining metal articles together, depending on the type of metal and the strength of the joint which is required.

连接金属件的方法有许多，其分类取决于需连接金属的类型和所要求焊点的强度。

句中 depending on the types of metals and the strength of the joint which is required 是现在分词短语做状语，起补充说明作用，而短语中 which is required 是定语从句，修饰 the joint；a number of 可译为“许多，一些”。

[2] One of the most important operations in soldering is that of cleaning the surfaces to be joined, this may be done by some acid cleaner.

钎焊中最重要的操作之一是清洁被焊接表面，该步骤可用含酸清洁剂来完成。

句中 that 指的是 operation；后面的 this 指的是 that of cleaning the surfaces to be joined。

[3] Flux is used to remove and prevent oxidation of the metal surfaces to be soldered, allowing the solder to flow freely and unite with the metal.

钎剂可用来去除和防止钎焊金属表面上的氧化物，从而使钎料可以自由流动并和金属结合。

句中 allowing the solder to flow freely and unite with the metal 是现在分词短语做状语，起补充说明作用。

Glossary of Terms

1. filler rod (electrode, covered electrode) 焊条
2. fusion welding 熔焊
3. pressure welding 压力焊
4. brazing (soldering) 铜焊，硬焊（软焊，锡焊，钎焊）
5. welding technique 焊接技术
6. welding technology (welding procedure) 焊接工艺
7. welding operation 焊接操作
8. welding sequence 焊接顺序
9. welding position 焊接位置
10. down-hand welding (flat position) 平焊
11. horizontal position welding 横焊
12. vertical position welding 立焊
13. overhead position welding 仰焊
14. arc welding 电弧焊
15. spot welding 点焊
16. lap welding 搭接焊
17. slot welding 槽焊
18. fillet welding 角焊
19. shielded metal arc welding (SMAW) 焊条电弧焊
20. gas metal arc welding (GMAW), metal inert gas (MIG) welding 熔化极气体保护电弧焊
21. flux-cored arc welding (FCAW) 药芯焊丝电弧焊
22. gas tungsten arc welding (GTAW), or tungsten inert gas (TIG) welding 钨极惰性气体保护焊

23. plasma arc welding (PAW) 等离子弧焊
24. submerged (shielded) arc welding (SAW) 埋弧焊
25. electroslag welding (ESW) 电渣焊
26. resistance welding (RW) 电阻焊
27. ultrasonic welding (USW) 超声波焊
28. friction welding (FRW) 摩擦焊
29. electron beam welding (EBW) 电子束焊
30. laser beam welding (LBW) 激光焊
31. oxyacetylene welding 氧乙炔焊
32. welding procedure specification 焊接工艺规程
33. manual arc welder 电焊工
34. weld inspector 焊接检验员
35. eye shield 护目镜
36. face shield （电焊用）面罩

Reading Materials

Welded Joints and Symbols

Common welded joints are illustrated in Figure 3-2-1; each has several elements. These are the types of joints, the types of welds, and the preparation for the weld. The elements can be put together in various ways. For example, a lap joint may be held by a fillet, plug, or slot weld, and a tee joint by a fillet or groove weld. The nature of the joint depends upon the kind and size of material, the process and the strength required. Material less than 0.25mm thick is usually lapped; thicker material is commonly butted. Butt joints are prepared for high-strength steels because they are more easily inspected and involve simpler stress patterns than lap joints. Lap joints are best for most pressure and resistance welding of sheets and for electron beam welding where no filler metal is added.

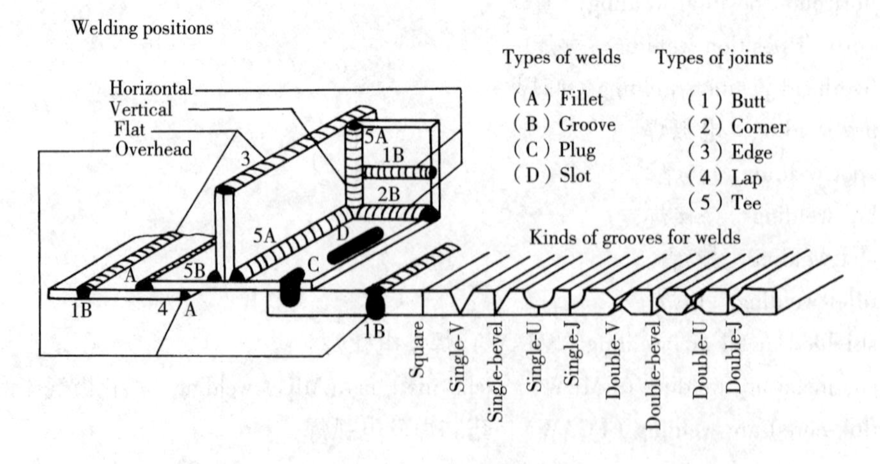

Figure 3-2-1 Types of welded joints

Proportions of welded joints have been standardized. Preferred sizes, dimensions, and charts for calculating the strengths and amounts of filler metal required for welded joints are given in reference texts and handbooks.

Position, as defined in Figure 3-2-1, is an important consideration for any welded joint. Gravity aids in putting down the weld metal into a flat position, which is the easiest and fastest to execute.

Precise instructions for any welded joint can be given on a drawing by a system of symbols and conventions. This is a special language and governed by definite rules, like the rules of grammar. A full list of symbols and their meanings is given in reference books. Several illustrations are presented in Figure 3-2-2.

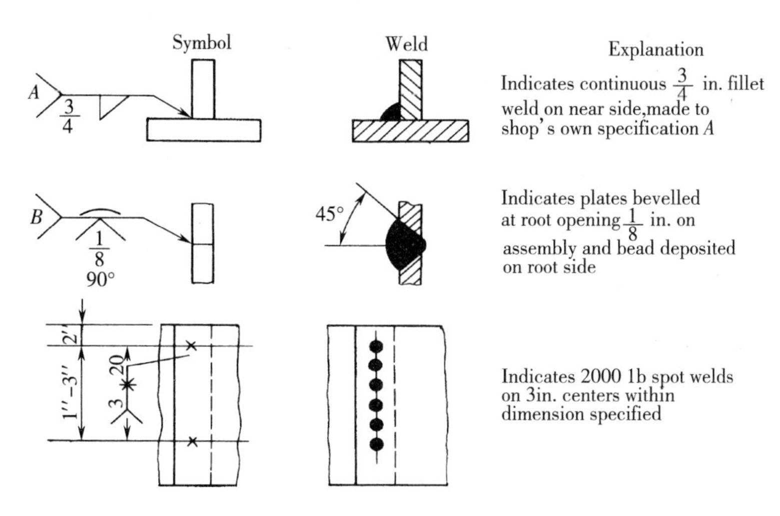

Figure 3-2-2 Examples of welding symbols

Arc Welding

Arc welding processes use a welding power supply to create and maintain an electric arc between an electrode and the base material to melt metals at the welding point. They can use either direct (DC) or alternating (AC) current, and consumable or non-consumable electrodes. The welding region is sometimes protected by some type of inert or semi-inert gas, known as a shielding gas, and filler material is sometimes used as well.

Shielded metal arc welding. One of the most common types of arc welding is shielded metal arc welding (SMAW), which is also known as manual metal arc welding (MMAW) or stick welding. Electric current is used to strike an arc between the base material and consumable electrode rod, which is made of steel and is covered with a flux that protects the weld area from oxidation and contamination by producing CO_2 gas during the welding process. The electrode core itself acts as filler material, making a separate filler unnecessary. The process is very versatile, requiring little operator training and inexpensive equipment. However, weld times are rather slow, since the consumable electrodes must be frequently replaced and because slag, the residue from the flux, must be chipped away after welding. Furthermore, the process is generally limited to welding ferrous materials, though speciality electrodes have made possible the welding of cast iron, nickel, aluminium, copper and other metals. The versatility of the method makes it popular in a number of applications, including repair work and construction. Figure 3-2-3 depicts the way in which metal is deposited from a shielded electrode. The typical setup with AC power source, is shown in Figure 3-2-4.

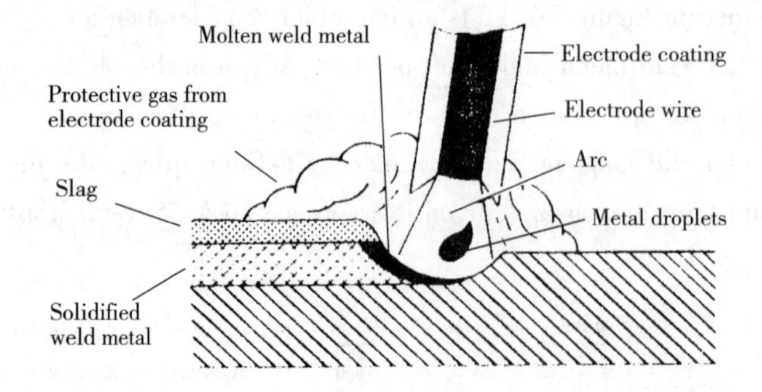

Figure 3-2-3 Schematic diagram of shielded metal arc welding

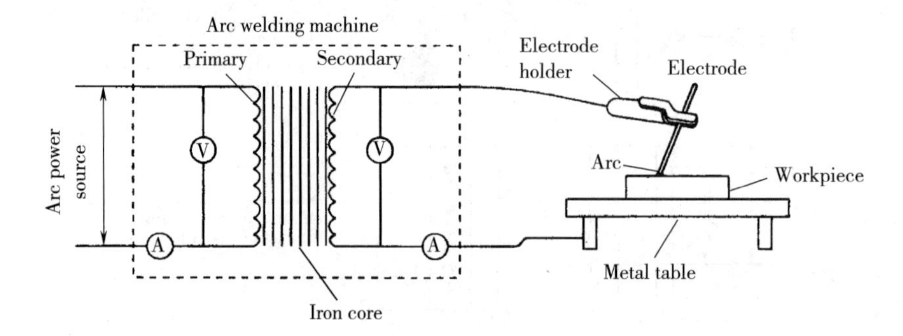

Figure 3-2-4 Sectional view of representation of the arc welding setup

Unit 3 Cold Working Processes

Text

Plastic deformation of metals below the *recrystallization* temperature is known as cold working, and is generally performed at room temperature. In some cases, however, the working may be done at mildly elevated temperatures to provide increased ductility and reduced strength. From a manufacturing viewpoint, cold working has a number of distinct advantages, and the various cold working processes have become extremely important①. *Significant* advances in recent years have extended the use of cold forming and the trend appears likely to continue.

When compared with hot working, the advantages of cold working include:

(1) No heating is required.

(2) Better surface finish is obtained.

(3) *Superior* dimension control.

(4) Better *reproducibility* and interchangeability of parts.

(5) Improved strength properties.

(6) Directional properties can be *imparted.*

(7) *Contamination* problems are minimized.

Some disadvantages associated with cold working processes include:

(1) Higher forces are required for deformation.

(2) Heavier and more powerful equipment is required.

(3) Less ductility is available.

(4) Metal surfaces must be clean and scale-free.

(5) Strain hardening occurs (may require *intermediate* anneals).

(6) Imparted directional properties may be *detrimental.*

(7) May produce undesirable residual stresses.

As a result, numerous cold working processes have been developed to perform a variety of deformations②. The major cold working operations can be classified basically under the headings of squeezing, bending, shearing and drawing as follows:

Squeezing: rolling *swaging*, cold forging, extrusion, sizing, *riveting*, coining, etc.

Bending: angle, roll draw and compression, *seaming flanging*, etc.

Shearing: blanking, piercing, notching, *shaving*, *trimming*, etc.

Drawing: bar and tube drawing, wire drawing, *spinning*, *embossing*, *stretch* forming, shell drawing, etc.

Questions

1. What is cold working?

2. What are some of the advantages of cold working? Disadvantages?
3. What are some of the advantages of cold deformation?
4. What are major cold working operations?
5. Describe what happens when sheet metal is sheared.

New Words and Expressions

1. recrystallization [riːˌkristəlaiˈzeiʃn] n. 再结晶
2. significant [sigˈnifikənt] adj. 重要的，有意义的
3. superior [sjuːˈpiəriə] adj. 优良的，较多的
4. reproducibility [riprəˌdjuːsəˈbiliti] n. 重复能力，再现性
5. impart [imˈpɑːt] vt. 给予（尤指抽象事物），传授，告知
6. contamination [kənˌtæmiˈneiʃən] n. 污染，污物
7. intermediate [ˌintəˈmiːdjət] adj. 中间的 n. 媒介，中间体
8. detrimental [ˌdetriˈmentl] adj. 有害的，伤害的
9. swaging [ˈsweidʒiŋ] n. 型锻，模锻
10. riveting [ˈrivitiŋ] n. 铆，铆接
11. seaming [ˈsiːmiŋ] n. 接缝，缝拢
12. flanging [ˈflændʒiŋ] n. 凸缘，卷边
13. shaving [ˈʃeiviŋ] n. 修整，剃齿，缺口修整加工
14. trimming [ˈtrimiŋ] n. 整修，切边
15. spinning [ˈspiniŋ] n. 旋转，旋压
16. embossing [imˈbɔsiŋ] n. 滚花，压纹
17. stretch [stretʃ] n. 拉深，拉紧
18. be known as 被称为，叫作，通称
19. associated with 使联合，结合

Notes

[1] From a manufacturing viewpoint, cold working has a number of distinct advantages, and the various cold working processes have become extremely important.

从制造的观点来看，冷作工艺有许多显著的优点，并且各种冷作工艺已成为极为重要的加工手段。

句中 from a manufacturing viewpoint 可译为“从制造的观点来看”。

[2] As a result, numerous cold working processes have been developed to perform a variety of deformations.

因此，众多冷作工艺已经用于进行金属的各种变形加工。

句子中 as a result 可译为“结果，因此，从而”，对照上下文，此处译为“因此”。

Glossary of Terms

1. cold working processes 冷作工艺
2. cold forming 冷成形
3. cold hardening (quenching) 冷作硬化，加工硬化
4. cold forging (hammering) 冷锻
5. cold forging drawing 冷锻件图
6. cold rolling 冷轧
7. cold heading 冷镦
8. cold extrusion 冷挤压
9. cold-draw steel 冷拔钢
10. cold-draw wire 冷拔钢丝
11. cold work tool steel 冷作工具钢
12. a punch press 冲床，冲孔机
13. mechanical working of metals 金属压力加工
14. thread rolling 滚轧螺纹法
15. rolling mill 轧钢机，轧钢厂
16. shaping roll 粗轧机
17. powder metallurgy (P/M) 粉末冶金
18. automotive components 汽车零件
19. piston rings 活塞环
20. super-plastic forming 超塑性成形
21. plastic deformation 塑性变形
22. room temperature 室温
23. work hardening 加工硬化
24. press work 压力加工

Reading Materials

Cold Rolling

Bars of all shapes, rods, sheets and strips are commonly finished in all common metals by cold rolling. Foil is made of the softer metals in this way. Cold-rolled sheets and strips make up an important part of total steel production and are major raw materials for some high-production consumer goods industries, such as for household appliances.

Metals are cold rolled for improved physical properties, good surface finish, textured surfaces, dimensional control, and machinability. Sheet steel that about 1.5mm thick is cold rolled as a matter of course because it cools too rapidly for practical hot rolling. Cold rolling produces uniform thicknesses and close tolerances in sheets and bars. Machinability of most steels is improved by cold working and for that reason cold-rolled or drawn stock is widely used in fast automatic machining operations.

Cold Extrusion

Cold extrusion, also called cold forming, cold forging and extrusion pressing, is normally done at room temperature. Its forms are like those of hot extrusion. Cold extrusion is done quickly, at ram speeds of 0.25 to 1.5 m/s, generates heat that raises the temperature several hundred degrees, and takes less force than if done slowly. Some parts are formed in one pressing in a single die; others in two or more stages in a series of dies, sometimes in conjunction with cold heading. Presses mostly used are the same as for sheet metal forming.

Several makes of continuous extruders have been developed to produce wire, rods, and various shapes from powders, ground-up scrap, or larger-diameter stock. They utilize one or more grooved wheels that take and compress the feedstock as they revolve and force it into the die through which the metal is extruded.

Advantages of cold extrusion are that it is fast, may improve physical properties and save heat treatment, wastes little or no material, can make parts with small radii and no draft, and can produce to small tolerances and save machining. Tolerances can be as close as $\pm 25\mu m$, but larger tolerances are much cheaper. Cold extrusion is competitive with deep drawing for making cups and deep shells. Extrusion has the advantage of requiring fewer steps thus saving tooling. Cold extrusion is competitive with casting and forging for some parts. Extrusions are usually lighter and stronger than castings. They don't need draft of flash to trim. They are not porous or brittle as castings may be. Tolerances are closer and less machining is required for extruded parts. Cheaper metals can sometimes be put in extrusions because the process improves the physical properties.

Processing of Metal Powders

In the manufacturing processes we have described thus far, the raw materials used have been metals and alloys either in a molten state (casting) or in solid form (metal working). This chapter describes the powder metallurgy (P/M) process, in which metal powders are compacted into desired and often complex shapes and sintered (heated without melting) to form a solid piece. This process first was used by the Egyptians in about 3000B. C. to make iron tools. One of its first modern uses was in the early 1900s to make the tungsten filaments for incandescent light bulbs. The availability of a wide range of metal-powder compositions, the ability to produce parts to net dimensions (net-shape forming), and the overall economics of the operation give this unique process its numerous attractive and expanding applications.

A wide range of parts and components are made by powder-metallurgy techniques: ①balls for ball-point pens; ②automotive components (which now constitute about 70% of the P/M market) such as piston rings, connecting rods, break pads, gears, cams and bushings; ③tool steels, tungsten carbides, and cermets as tool and die materials; ④graphite brushes impregnated with copper for electric motors; ⑤magnetic materials; ⑥metal filters and oil-impregnated bearings with controlled porosity; ⑦metal foams; ⑧surgical implants, and several others for aerospace, nuclear and industrial applications. Advances in this technology now permit structural parts of aircraft, such as landing gear components, engine-mount supports, engine disks, impellers, and engine nacelle frames to be made by powder metallurgy.

P/M has become competitive with processes (such as casting, forging and machining), particularly for relatively complex parts made of high-strength and hard alloys.

The most commonly used metals in P/M are iron, copper, aluminum, tin, nickel, titanium and the refractory metals.

Unit 4 Hot Working Processes

Text

The metal working processes are traditionally divided into hot working and cold working processes. The division is on the basis of the amount of heat applied to the metal before applying the mechanical force.

Metals are worked by pressure in the *primary* processes for two reasons: to form desired shapes, and to improve physical properties. The results depend on whether the work is done hot or cold. Hot working is done above the recrystallization temperature. It is at or near room temperature for lead, tin and zinc. It is above the critical temperature for steel, as depicted in Figure 3-4-1.

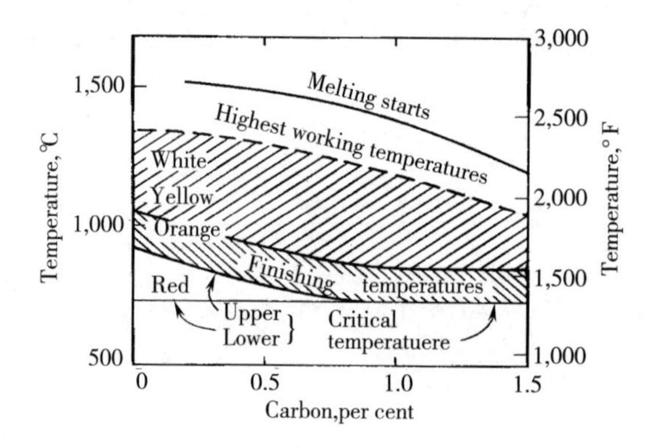

Figure 3-4-1 Range of rolling and forging temperatures for carbon steel

Hot working. The properties of metal are different when it is above and below its recrystallization temperature. The strength of a metal decreases as temperature rises, and its grains can be *distorted* more easily. If a ductile crystal is distorted by cold working, it does not visibly come apart, but its lattice structure is *fragmented*①. New and smaller crystals form out of the fragments. If the temperature is dropped soon, a fine structure results. However, if the metal is held above the recrystallization temperature and below the melting point, its crystals grow larger. Small crystals tend to combine, and large ones absorb small ones. The higher the temperature, the faster the growth②. The longer the time, the larger the grains become. These conditions help explain the following advantages of pressing or working hot metals.

(1) True hot working does not change the hardness or ductility of the metal. Grains distorted and strained during the process soon change into new undeformed grains.

(2) The metal is made *tougher* because the grains are reformed into smaller and more numerous crystals.

(3) The metal is made tougher because its *pores* are closed and *impurities* are *segregated.* Slag and other inclusions are squeezed into fibers with a definite orientation. Metal is hot worked to orient

the flow lines as nearly as possible for strength in the direction of largest stress.

(4) Less force is required, the process is faster, and smaller machines can be used for a given amount of heat as compared with cold working because the metal is weaker.

(5) A metal can be pushed into *extreme* shapes when hot without *ruptures* and *tears* because the crystals are more pliable and continually reformed.

Hot working is done well above the critical temperature to gain most of the benefits of the process but not at a temperature high enough to promote extreme grain coarsening③.

Hot working has several major disadvantages. It requires heat-resistant tools which are relatively expensive. The high temperatures oxidize and form scale on the surface of the metal. Close tolerances cannot be held. Cold working is necessary to overcome these *deficiencies*.

Some of the hot working processes that are of major importance in modern manufacturing are: rolling, forging (smith, drop, press, upset), extrusion, etc.

Questions

1. How does a metal act above its recrystallization temperature?
2. What are the advantages of hot working metal?
3. What are the disadvantages of hot working metal?
4. Why does a deformed metal want to recrystallize?

New Words and Expressions

1. primary [ˈpraiməri] adj. 第一位的，最初的
2. distort [disˈtɔːt] vt. 变形；歪曲，使……失真
3. fragmented [frægˈmentid] adj. 成碎片的，片断的
4. tough [tʌf] adj. 不易磨损的，紧韧的
5. pore [pɔː] n. 孔，孔隙
6. impurities [imˈpjuəriti] n. 不纯，杂质
7. segregate [ˈsegrigeit] adj. 偏析的
8. extreme [iksˈtriːm] n. 极限值 adj. 极端的
9. rupture [ˈrʌptʃə(r)] n. 破裂，断裂
10. tear [tiə] n. 撕
11. deficiency [diˈfiʃənsi] n. 缺乏，缺陷
12. on the basis of 在……基础上，根据
13. applied to 施加，应用
14. be squeezed into 被挤压成……

Notes

[1] If a ductile crystal is distorted by cold working, it does not visibly come apart, but its lattice structure is fragmented.

塑性晶体如果在冷作加工作用下发生变形，虽然它没有发生明显破碎，但它的晶格结构发生了分裂。

句中“if a ductile crystal is distorted by working”为由 if 引导的条件状语从句，可译为“塑性晶体如果在加工作用下发生变形”。

[2] The higher the temperature, the faster the growth.

(加热) 温度越高，(金属的晶粒) 成长越快。

本句为“the + 比较级……, the + 比较级……”结构，句子中省略了 be 动词。

[3] Hot working is done well above the critical temperature to gain most of the benefits of the process but not at a temperature high enough to promote extreme grain coarsening.

热作加工在高于金属临界温度的条件下进行，以最大程度地发挥该工艺的优势，但温度不宜过高以避免金属晶粒粗化的加剧。

句中“the benefits of ”可译为“……的优势”；“... enough to ... ”意为“……足够……，足以……”。

Glossary of Terms

1. plastic working of metal 金属塑性加工
2. hot working processes (technology) 热作工艺
3. recrystallization temperature 再结晶温度
4. critical temperature 临界温度
5. forging temperature interval 锻造温度范围
6. heating temperature 加热温度
7. hot forging 热锻
8. warm forging 温锻
9. hot forging drawing 热锻件图
10. draft angle 模锻斜度
11. open die forging 自由锻，开式模锻
12. loose tooling forging 胎模锻
13. closed die forging 闭式模锻
14. no-flash (flashless) die forging 无飞边模锻
15. precision forging 精密锻造
16. finish forging 终锻
17. hammer forging 锤锻
18. upset forging 镦锻
19. drop hammer (drop stamp) 落锤

20. drawing out 拔长
21. hot extrusion 热挤压
22. forward (direct) extrusion 正挤压
23. backward (indirect) extrusion 反挤压
24. combined extrusion 复合挤压
25. sideways (lateral) extrusion 侧向挤压
26. continuous extrusion 连续挤压
27. hydraulic press 液压机
28. mechanical press 机械压力机
29. gap-frame press 开式压力机
30. straight side press 闭式压力机
31. crank press 曲柄压力机
32. single action press 单动压力机
33. double action press 双动压力机
34. friction screw press 摩擦压力机

Reading Materials

Rolling

Rolling is a process where the metal is compressed between two rotating rolls for reducing its cross sectional area, as depicted in Figure 3-4-2. It is the first step in converting cast material into finished wrought products. Many finished parts, such as hot rolled structural shapes, are completed entirely by hot rolling. More often, however, hot rolled products, such as sheets, plates, bars and strips, serve as input material for other processing, such as cold forming or machining. This is one of the most widely used of all the metal working processes, because of its higher productivity and low cost. Rolling would be able to produce components having constant cross-section throughout its length, but not very complex shapes. It is also possible to produce special sections such as railway wagon wheels by rolling individual pieces.

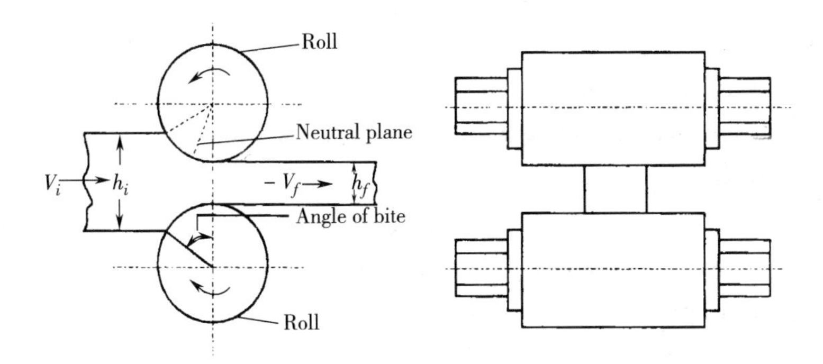

Figure 3-4-2 Schematic of the rolling process

Principles of metal rolling. When metal is rolled, it passes and is squeezed between two revolving rolls in the manner indicated in Figure 3-4-3. The crystals are elongated in the direction of rolling, and the material emerges at a faster rate than it enters. In hot rolling the crystals start to reform after leaving the zone of stress, but in cold rolling they retain substantially the shape given by the action of the rolls.

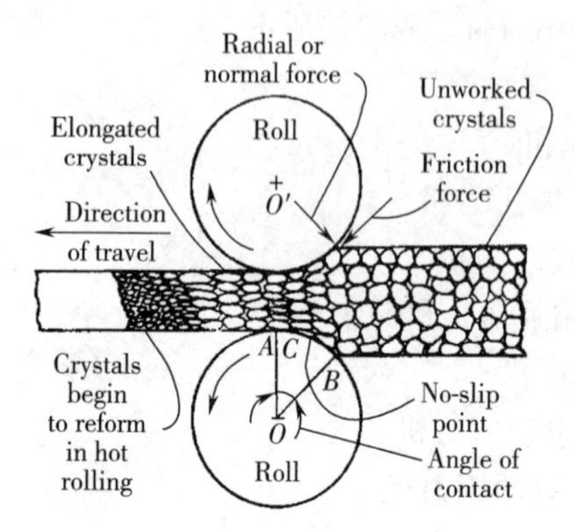

Figure 3-4-3 Sketch to show what happens when metal is rolled

Rolling temperatures. Hot rolling usually is completed about 50℃ to 100℃ above the recrystallization temperature. Maintenance of such a finishing temperature assures the production of a uniform fine grain size and prevents the possibility of unwanted stain hardening.

Rolling mill stands are available in a variety of roll configurations. Early reductions, often called primary roughing or breakdown passes, usually employ a two-high or three-high configuration with 600mm to 1400mm diameter rolls.

Characteristics, quality, and tolerances of hot rolling products. Because they are rolled and finished above the recrystallization temperature, hot rolled products have minimum directional properties and are relatively free of residual stresses. These characteristics, however, often depend on the thickness of the product and the existence of complex sections that may prevent either uniform working in all directions or uniform cooling. Thin sheets show some definite directional characteristics, whereas thicker plate, for example above 20mm, usually will have very little. A complex shape, such as an I or H-beam, will warp substantially if a portion of one flange is cut away because of the residual stress in the edges.

The surfaces of hot rolled products are, of course, slightly rough and covered with a tenacious high-temperature oxide known as mill scale. However, with modern procedures, surprisingly smooth surfaces can be obtained.

The variety of hot rolled products is considerable, despite the kind of metal and the size of the product. For most products produced in reasonably large tonnages, the tolerance is from 2% to 5% of the size (height or width).

Forging

Forging is the plastic working of metal by means of localized compressive forces exerted by manual or power hammers, presses, or special forging machines. It may be done either hot or cold. However, when it is done cold, special names usually are given to the processes. Consequently, the term "forging" usually implies hot forging done above the recrystallization temperature.

Forging may be done in open or closed dies. Open die forgings are nominally struck between two flat surfaces, but in practice the dies are sometimes vee shaped, half round, or half oval. Closed die forgings are formed in die cavities. All forging takes skill, but more is required with open than with closed dies. Faster output and smaller tolerances are obtained with a closed die. Open dies are, of course, much less costly than closed dies and more economical for a few parts. Either open or closed die forging may be done on most hammers and presses.

Various forging processes have been developed to provide great flexibility, making it economically possible to forging a single piece or to mass produce thousands of identical parts. The metal may be ①drawn out, increasing its length and decreasing its cross section; ②upset, increasing the cross section and decreasing the length; or ③squeezed in closed impression dies to produce multidirectional flow.

The common forging processes are:

(1) Open-die hammer or smith forging. This is the traditional forging operation done openly or in open dies by the village blacksmith or modern shop floor by manual hammering or by power hammers.

(2) Impression-die drop forging. Drop forging is the name given to the operation of forging parts hot on a drop hammer with impression or cavity dies. The products are known as drop forgings, closed die forgings, or impression die forgings. They are made from carbon and alloy steels and alloys of aluminum, copper, magnesium, nickel and other metals.

(3) Press forging. Similar to drop forging, the press forging is also done in closed impression dies with the exception that the force is a continuous squeezing type applied by the hydraulic presses.

(4) Upset forging. Unlike the drop or press forging where the material is drawn out, in upset forging, the material is only upset to get the desired shape.

Quality and cost. The dimensions of series of forgings from a die vary because of differences in the behavior of the material, temperatures, closing of the die, mismatch of the die halves, and enlargement of the cavities as they wear. A tolerance of ±0.8mm is considered good for small carbon steel forgings and may be as large as 7mm in all directions for large pieces. Tolerances of 0.25mm and less have been held on precision press forgings, but at higher cost.

Many forgings are finished by machining to close tolerances and must have enough stock on the surfaces to be machined. The least stock is about 1.5mm per surface on small forgings, and may be as much as 7mm or more on large ones.

Parts made from forgings may be made also as castings, cut from standard shapes, fabricated by welding pieces together, or made in other ways. A forging can be made stronger, more shock and fatigue resistant, and more durable than other forms.

Unit 5 Blanking Technique

Text

In the following discussion, certain die *terminology* will be used *frequently*. Figure 3-5-1 presents the terms most commonly *encountered*.

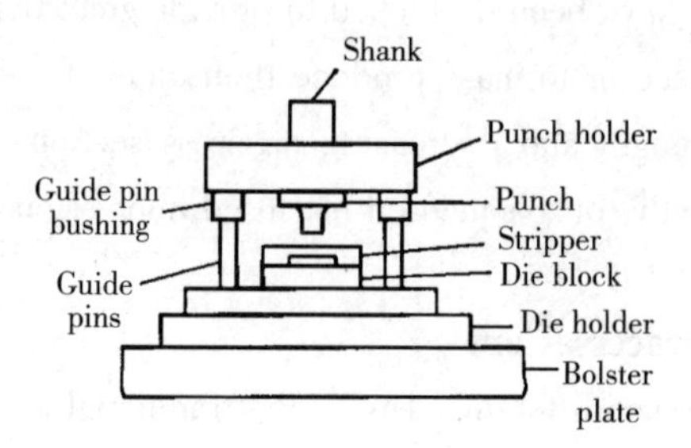

Figure 3-5-1 Common components of a simple die

Shear action in die cutting operations. The cutting of metal between die components is a shearing process in which the metal is *stressed* in shear between two cutting *edges* to the point of *fracture*, or beyond its ultimate strength.

The metal is subjected to both tensile and compressive stresses (Figure 3-5-2); *stretching* beyond the elastic limit occurs, then plastic deformation, reduction in area, and finally, fracturing starts through cleavage planes in the reduced area and becomes complete.

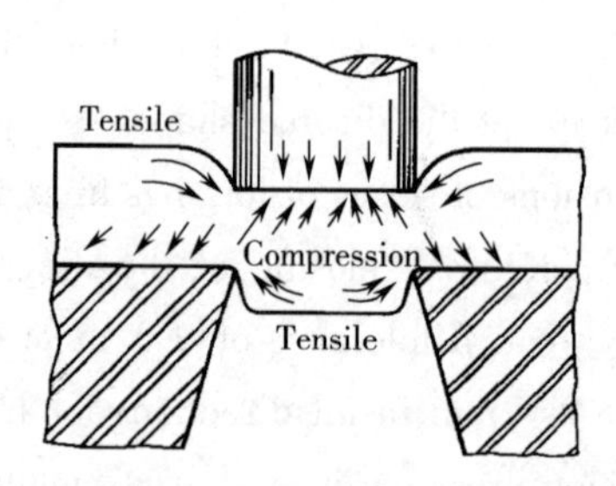

Figure 3-5-2 Stresses in die cutting

The fundamental steps in shearing or cutting are shown in Figure 3-5-3. The pressure applied by the punch on the metal tends to deform it into the die opening. When the elastic limit is exceeded by further loading, a portion of the metal will be forced into the die opening in the form of an *embossed* pad on the lower face of the material. A *corresponding* depression results on the upper face, as indicated in Figure 3-5-3a. As the load is further increased, the punch will penetrate the metal to a certain depth and force an equal portion of metal thickness into the die, as indicated in Figure 3-5-3b. This *penetration* occurs before fracturing starts and reduces the cross-sectional area of metal through which the cut is being made. Fracture will start in the reduced area at both upper and lower cutting edges, as indicated in Figure 3-5-3c. If the *clearance* is suitable for the material being cut, these fractures will spread toward each other and *eventually* meet, causing complete *separation*①. Further travel of the punch will carry the cut portion through the *stock* and into the die opening.

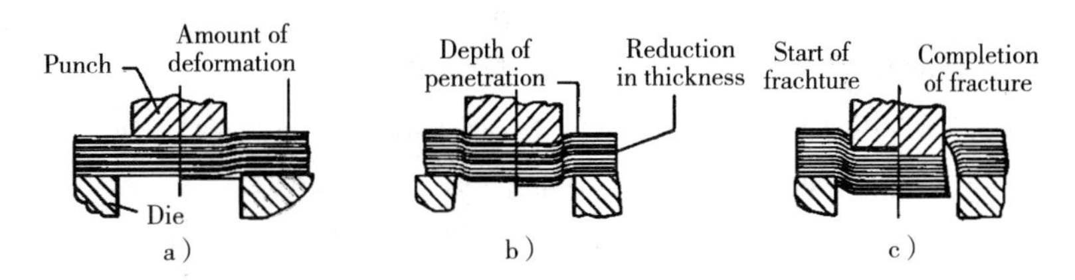

Figure 3-5-3 Steps in shearing metal

Center of pressure. If the *contour* to be blanked is irregularly shaped, the *summation* of shearing forces on one side of the center of the ram may greatly exceed the forces on the other side. Such irregularity results in a bending moment in the press ram, and *undesirable deflections* and *misalignment.* It is therefore necessary to find a point about which the summation of shearing forces will be *symmetrical.* This point is called the center of pressure, and is the center of gravity of the line that is the *perimeter* of the blank. It is not the center of gravity of the area.

The press tool will be designed so that the center of pressure will be on the *axis* of the press ram when the tool is mounted in the press②.

Questions

1. What is the center of pressure?
2. Describe shear action in die cutting operations.
3. What are the fundamental steps in shearing or cutting?
4. What is necessary for proper clearance in shearing or cutting?
5. When the contour to be blanked is irregularly shaped, why must the center of pressure be calculated?

New Words and Expressions

1. terminology [ˌtəːmiˈnɔlədʒi] n. 术语，专门名词
2. frequent [ˈfriːkwənt] adj. 频繁的 vt. 常去，常到 ~ly adv. 常常
3. encounter [inˈkauntə] vt. 遭遇，碰撞 vi. 偶遇 n. 遭遇
4. stress [stres] n. 压力，应力 vt. 强调，使受应力
5. edge [edʒ] n. 刃，刀口 vi. 使锐利，给……镶边
6. fracture [ˈfræktʃə] n. 断口，断裂 vt. 断裂
7. stretch [stretʃ] vt. & vi. 伸展，展开，加宽
8. emboss [imˈbɔs] vt. 拷花，压纹，在……上浮雕图案
9. correspond [kɔrisˈpɔnd] vi. 相当，对应，符合
10. penetration [peniˈtreiʃən] n. 渗透，穿透能力

11. clearance [ˈkliərəns] n. 清除，间隙，空隙
12. eventual [iˈventjuəl] adj. 最后的，可能发生的
13. separation [sepəˈreiʃən] n. 分离，分类，间隔
14. stock [stɔk] n. 托柄，原料，座 vt. 给……装上把手
15. contour [ˈkɔntuə] n. 轮廓，外形 vt. 描画轮廓线
16. summation [sʌˈmeiʃən] n. 总结，总数，加法
17. undesirable [ˌʌndiˈzaiərəbl] adj. 不希望的，不合乎需要的
18. deflection [diˈflekʃən] n. 偏移，偏差，挠曲
19. misalignment [ˈmisəlainmənt] n. 未对准，失调，不重合
20. symmetrical [siˈmetrikəl] adj. 对称的，匀称的
21. perimeter [pəˈrimitə] n. 圆周，周长
22. gravity [ˈgræviti] n. 重力，重要性
23. axis [ˈæksis] n. 轴，轴线，坐标轴
24. be subjected to 易受……的，须经……的
25. in the form of 以……的形式

Notes

[1] If the clearance is suitable for the material being cut, these fractures will spread toward each other and eventually meet, causing complete separation.

对于被剪切的材料，如果间隙适当，则裂纹将相向扩展并最终相遇，从而实现材料的完全分离。

原文主句中有两个谓语动词，一个是 spread，另一个是 meet，它们共用一个主语，即 these fractures；分词短语 causing complete separation 用作状语，表示结果，可译为“从而实现材料的完全分离”。注意：在科技文章中，分词短语做状语时，一般都用“,”把它与句子的其他部分分开。

[2] The press tool will be designed so that the center of pressure will be on the axis of the press ram when the tool is mounted in the press.

冲压模具的设计应满足当把模具安装在压力机上时，模具的压力中心位于压力机滑块的轴线上。

“so that”可引导目的和结果状语从句，前者意为“为了，以便”，后者意为“以致，使得，因此”；在本句中“so that”即为目的状语从句。

Glossary of Terms

1. blanking die 冲裁模
2. blanking force 冲裁力
3. piercing die 冲孔模
4. die, stamping and punching die 冲模
5. die life 冲模寿命

6. die shut height 模具闭合高度
7. shut height of press machine 压力机闭合高度
8. clearance between punch and die 凸凹模间隙
9. shearing force diagram 剪力图
10. peak die load 模具最大负荷
11. outer slide 外滑块
12. center of die, center of load 压力中心
13. clamp plate (ring) 压板(夹紧环)
14. shearing force (plane) 剪切力(平面)
15. side clearance angle 侧隙角
16. side locating face 侧定位面
17. side-push plate 侧压板
18. shuttle table 移动工作台
19. matrix plate 凹模固定板
20. material removal rate 材料切除率
21. sheet forming 板料成形,冲压
22. forming die 成形模
23. bed die 下模,底模
24. composite die, combined die 组合模,拼合模
25. compound dies 复合模
26. compound blank and pierce dies 落料冲孔模
27. shaving die 切边模,修边模
28. shankless die 无柄模具
29. scrapless progressive die 无废料连续模
30. return-blank type blanking die 顶出式落料模
31. restriking die 矫正模,校平模,整修模
32. reducing die 缩口模,缩径模
33. cutting-off die 切断模
34. bending die 弯曲模
35. springback 回弹
36. drawing die 拉深模

Reading Materials

Punching Die

The structural design of the die affects the workpiece quality as well as the production cost directly. Therefore, the investigation on the structure and characteristics of the punching die is crucial to realize the blanking production and develope the blanking technique.

The variety of the workpieces results in the variety of the structures of the punching dies. The blanking dies can be classified according to different features.

(1) Classification according to the process property. It can be classified as punching, blanking, trimming, cut-off, parting, lancing, shaving and fine blanking dies.

(2) Classification according to the process combination. It can be classified as single (simple), progressive (continuous) and compound dies.

(3) Classification according to the guiding pattern between the punch and die. It can be classified as guideless die and dies with guide plate, guide pillar and guide tube.

(4) Classification according to the stripping equipment. It can be classified as dies with fixed stripper and elastic stripper.

(5) Classification according to the stop gauge. It can be classified as dies with fixed stop pin, moved stop pin, pilot and pilot punch.

(6) Classification according to the material of the punch and die. It can be classified as carbide, steel bonded carbide, steel strip, rubber pad and polyurethane pad dies.

Typical Structure of Blanking Die

Simple die. The die that only one process is carried out in one press stroke is called simple die. Its structure is simple (Figure 3-5-4), so it can be easily manufactured. It is applicable to small batch production.

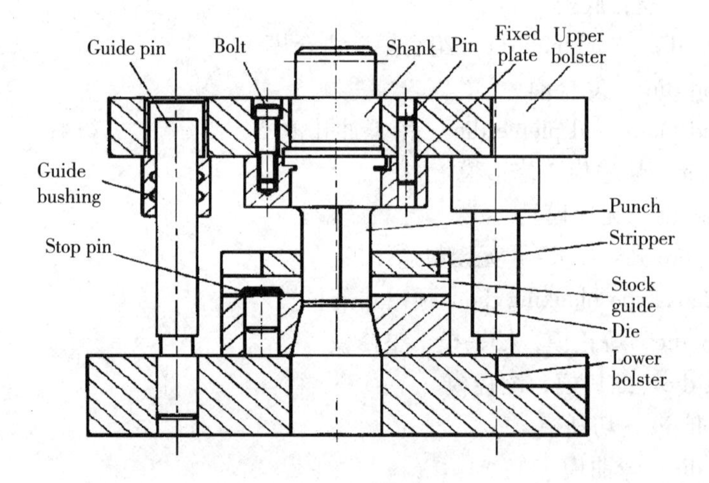

Figure 3-5-4 Simple die

Progressive die. The die that several blanking processes are carried out at different positions of the die in one press stroke is called progressive die, as shown in Figure 3-5-5. In the operation, the locating pin 2 aims at the locating holes punched previously, and the punch moves downwards to punch by punch 7 and to blank by punch 1, thus the workpiece 8 is produced. When the punch returns, the stripper 6 scrapes the blank 5 from the punch die, the blank 5 moves forward one step and then the second blanking begins. Above steps are repeated continually. The step distance of the blank is controlled by a stop pin.

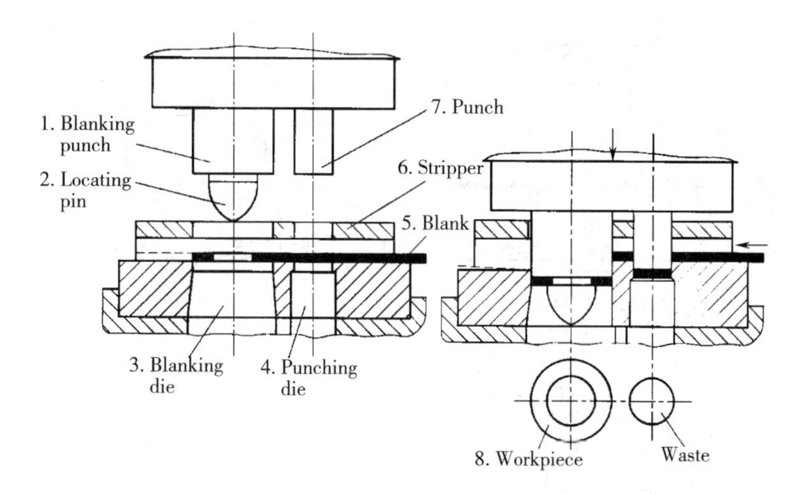

Figure 3-5-5 Progressive die for blanking and punching

Compound die. The die that several processes are carried out at the same die position in one press stroke is called compound die, as shown in Figure 3-5-6. The main characteristic of the compound die is that the part 1 is both the punch and the die. The outside circle of the punch die 1 is the cutting edge of the blanking punch, while the inside hole is a deep drawing die. When the slide moves downwards along with the punch die 1, the blanking process is done first by the punch die 1 and the blanking die 3, the blanked workpiece is pushed by deep drawing punch 7, and then the deep drawing die moves downwards to carry out deep drawing operation. The ejector 6 and the stripper 4 push the deep drawn workpiece 9 out of the die when the slide returns. The compound die is suitable for mass production and high accuracy blanking.

According to the preceding analysis on the types of the die structure, the die parts can be classified into two categories by its function.

(1) Technological structure parts.

These parts take part in the performance of the technological process and contact with the blank directly, which include the working parts (performing the stamping forming directly); locating parts (guaranteeing an excellent guide for the blank feeding, controlling the feed distance of the blank, obtaining a correct position of the blank); stripping, holding and ejecting parts (pressing the blank, knocking out the workpiece or waste from the die).

(2) Assistant structure parts.

These parts neither take part in the technological process nor contact with the blank directly, which include the guiding parts (guaranteeing the correct relative position between the punch and die to improve the quality of the blanking workpiece); the supporting and holding parts (installing the die to the press for transferring the working pressure); and the fastening parts (connecting and fastening separate parts of the die), etc.

The classification of the die is shown in the Table 3-5-1.

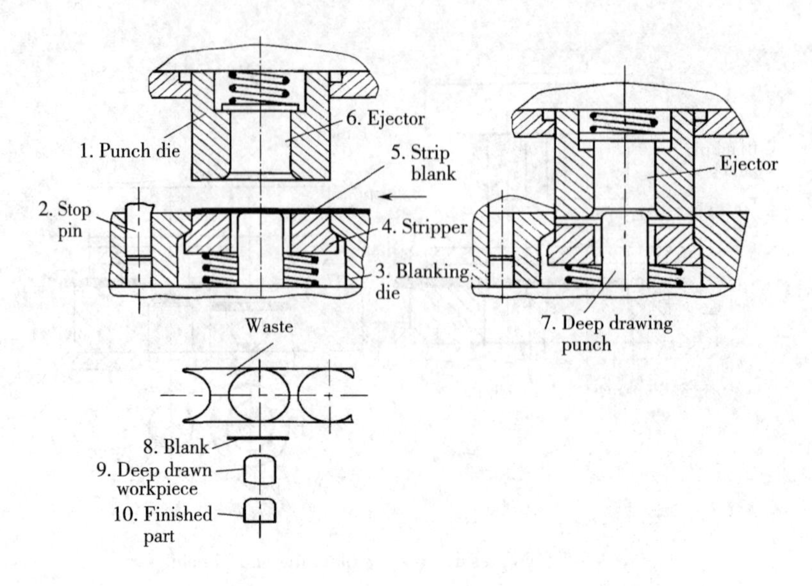

Figure 3-5-6 Compound die for blanking and deep drawing

Table 3-5-1 Classfication of the compound die

Number	Classification		Name of parts and components
1	Technological structure parts	Working parts	Punch, die, punch die
		Locating parts	Stop pin and pilot, stock guide (guide rule), locating pin (locating plate), side guide, kicker
		Stripping, holding and ejecting parts	Stripper, blank holder, ejector, knock out
2	Assistant structure parts	Guiding parts	Pillar, guide bushing, guide plate, guide tube
		Supporting and holding pars	Upper and lower bolster, fastening plate of the punch and die, shank, bolster plate, limiter
		Fastening and other parts	Bolt, pin, others

Unit 6 Plastic Processing

Text

There are two main steps in the manufacture of plastic products. The first is a chemical process to create the resin. The second is to mix and shape all the material into the finished article or product.

Plastic objects are formed by *compression*, *transfer* and *injection* molding. Other processes are casting, extrusion, *laminating*, *filament winding*, sheet forming, joining, foaming and machining. Some of these and still others are used for rubber. A reason for a variety of processes is that different materials must be worked in different ways. Also, each method is advantageous for certain kinds of products. The principles of operation and *merits* of the processes will be discussed.

Compression molding. In compression molding a proper amount of material in a cavity of a mold is squeezed by a punch, also called a force. The plastic is heated in most cases between 120℃ and 260℃ (250 ℉ and 500 ℉), softens, and flows to fill the space between the force and the mold. The mold is kept closed for enough time to permit the formed piece to harden. This is done in a press capable of exerting 15 MPa to 55 MPa (2000 psi to 8000 psi) over the area of the work projected on a plane normal to the ram movement, depending on the design of the part and the material.

Compression molding is mostly for thermosetting plastics which have to be cured by heat in the mold. Other methods are faster for large-quantity production of thermoplastics. Loose molding compound may be fed into a mold, but a cold-pressed tablet or rough shape, called a preform, may be prepared for more rapid production. For efficient heat transfer, parts should be simple with walls uniform and preferably not over 3mm (1/8 in) thick. Even so, it may take several minutes to heat and cure a charge. This time may be reduced as much as 50% by preheating the charge. To speed the process as much as possible, molding presses are usually semi-or fully automatic.

The three basic types of compression molds for plastics are shown in Figure 3-6-1. The force fits *snugly* in the positive-type mold. The full pressure of the force is exerted to make the material fill out the mold. The amount of charge must be controlled closely to produce a part of accurate size.

The force is a close fit in a semi-positive type mold only within the last millimeter of travel. Full pressure is exerted at the final closing of the mold, but excess material can escape, and the charge does not have to be controlled so closely. This type is considered best for large-quantity production of pieces.

The force does not fit closely but closes a flash or overflow type mold by bearing on a narrow flash *ridge* or cut-off area. The amount of material does not need to be controlled closely, and the excess is squeezed out around the cavity in a thin flash. Some material is wasted, and all pieces must be *trimmed*. Full pressure is not impressed on the workpiece. A mold of this kind is usually cheapest to make.

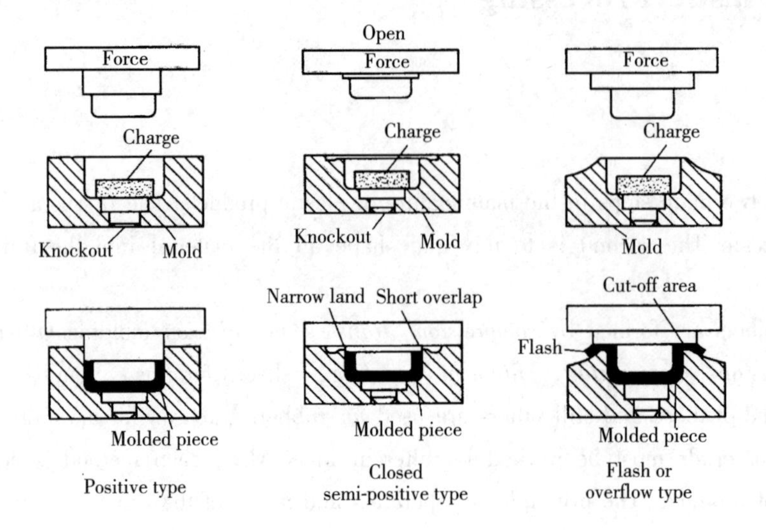

Figure 3-6-1 Three types of compression molds for plastics

Transfer molding. In transfer molding, also called an extrusion or gate molding, the material is heated and compressed in one *chamber* and forced through a sprue, runner, and *orifice* into the mold cavity① (Figure 3-6-2). The mold is costly, but closer tolerances and more uniform density can be held, time is generally shorter for thick sections, and thick and thin sections and inserts can be molded with less trouble than in other molds. The reason is that the material enters the mold under pressures of 40 to 85 MPa (6000 to 12000 psi) and acts like a fluid.

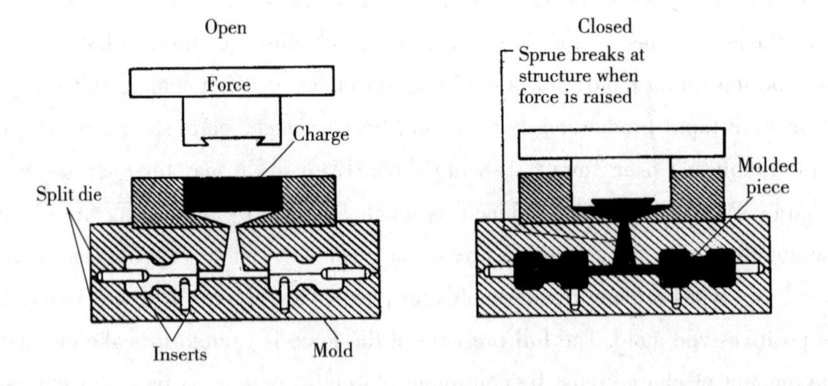

Figure 3-6-2 Transfer mold

Injection molding. Systems of injection molding plastic material are sketched in Figure 3-6-3. The oldest is the single-stage *plunger* method. When the plunger is drawn back, raw material falls from the *hopper* into the chamber. The plunger is driven forward to force the material through the heating cylinder where it is softened and *squirted* under pressure into the mold②. The single-stage *reciprocating* screw system has become more popular because it prepares the material more thoroughly for the mold and is generally faster. As the screw turns, it is pushed backward and crams the charge from the hopper into the heating cylinder. When enough material has been prepared, the screw stops turning and is driven forward as a plunger to ram the charge into the die. In a two-stage system, the material is plasticized in one cylinder, and a definite amount transferred by a plunger or screw into a shot chamber from which a plunger injects it into the mold.

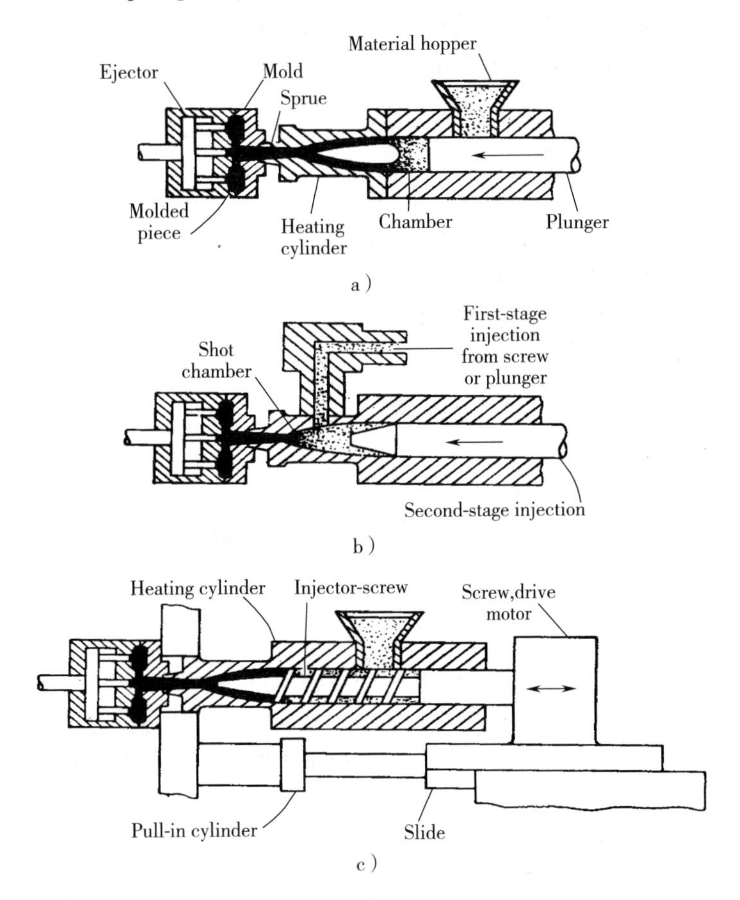

Figure 3-6-3 Injection molding systems

a) Conventional single-stage plunger type b) Two-stage plunger or screw-plasticisor types

c) Single-stage reciprocating screw type

Questions

1. Describe compression molding of plastics and the three basic types of molds used.
2. Describe the flash type (overflow type) mold.
3. How is injection molding done, and what are its advantages and disadvantages?
4. What difficulties can be overcome by transfer molding?

New Words and Expressions

1. compression [kəmˈpreʃ(ə)n] n. 压缩，压力

2. transfer [træns'fəː] vt. 转 [迁] 移，传送 ['trænsfəː] n. 转移，迁移
3. injection [in'dʒekʃən] n. 注射，加入，喷射（油）
4. laminating ['læmineitiŋ] n. 层压（法），层合（法）
5. filament ['filəmənt] n. 细丝，灯丝
6. winding ['waindiŋ] n. 绕组，绕法 adj. 卷绕的
7. merit ['merit] n. 长处，优点；价值
8. snugly ['snʌgli] adv. 舒适地
9. ridge [ridʒ] n. 脊；隆起物
10. trimmed [trimd] adj. 修整过的
11. chamber ['tʃeimbə] n. 室，腔
12. orifice ['ɔrifis] n. 孔；喷嘴
13. plunger ['plʌndʒə] n. 柱塞；冲杆
14. hopper ['hɔpə] n. 漏斗，加料斗
15. squirt [skwəːt] v. 喷出 n. 喷射器
16. reciprocate [ri'siprəkeit] v. 往复运动，互换
17. be driven forward to 向前施加压力
18. be prepared for（to do）sth. 对（做）某事有所准备

Notes

[1] In transfer molding, also called an extrusion or gate molding, the material is heated and compressed in one chamber and forced through a sprue, runner, and orifice into the mold cavity.

传递模也被称作挤压模或浇口模，塑料在加料室中被加热并压缩，然后通过直浇道、横浇道和浇注系统（喷嘴）进入模具型腔。

句中 also called an extrusion or gate molding 为同位语，进一步说明 transfer molding。

[2] The plunger is driven forward to force the material through the heating cylinder where it is softened and squirted under pressure into the mold.

柱塞向前施加压力使塑料通过加热筒，塑料在加热筒中被软化并且在压力作用下喷入到模具型腔中。

句中“is driven forward to ...”可译为“向前施加压力”；“where it is softened and squirted under pressure into the mold”为定语从句，修饰“heating cylinder”。

Glossary of Terms

1. injection mold 注射模
2. transfer mold 压注模（也称传递模）
3. portable transfer mold 移动式传递模
4. fixed transfer mold 固定式传递模
5. compression mold 压缩模（也称压制模）
6. flash type mold 溢式压缩模

7. semi-positive mold 半溢式压缩模
8. positive mold 不溢式压缩模
9. parting surface 分型面
10. runner less mold 无流道模
11. hot runner mold 热流道模
12. warm runner mold 温流道模
13. ring gate 环形浇道
14. pin-point gate 点浇道
15. edge gate 侧浇道
16. mold opening force 开模力
17. sprue puller 拉料杆
18. sprue bush 浇口套
19. injection speed 注射速度
20. phenolics（PF） 酚醛塑料
21. epoxide（EP） 环氧塑料
22. epoxy resin 环氧树脂
23. polyethylene（PE） 聚乙烯
24. polystyrene（PS） 聚苯乙烯
25. polypropylene（PP） 聚丙烯
26. polyamide（nylon）（PA） 聚酰胺，尼龙
27. polycarbonate（PC） 聚碳酸酯
28. polyvinyl chloride（PVC） 聚氯乙烯

Reading Materials

Properties of Plastics and Their Types

Plastics. It is difficult to give a precise definition of the term plastics. Basically, it covers a group of materials characterized by large molecules that built up by joining small molecules, usually artificially. Practically, it is sufficient to say that they are natural or synthetic resins, or their compounds, that can be molded, extruded, cast, or used as films or coatings. Most of them are organic substances, usually containing hydrogen, oxygen, carbon and nitrogen.

Properties of plastics. Because there are so many plastics with new ones becoming available almost continuously, it is helpful to have a knowledge of the general properties which they possess and the properties of the several basic types.

（1） Light weight. Most plastics have specific densities between 1.1g/cm^3 and 1.6g/cm^3, compared with about 1.75g/cm^3 for magnesium. Thus they are the lightest of the engineering materials.

（2） Corrosion resistance. Many plastics perform well in hostile, corrosive environments.

（3） Electrical resistance. They are widely used as insulating materials.

(4) Low thermal conductivity. They are relatively good heat insulators.

(5) Wide range of colors, transparent, or opaque. Many plastics have an almost unlimited color range and the color goes throughout, not just on the surface.

(6) Surface finish. Excellent surface finishes can be obtained by processes used to convert the raw material to the final shape. No added operations are required.

(7) Comparatively low cost.

Types of plastics. Phenolics (PF): oldest of the plastics, but still widely used; hard, relatively strong, low cost, and easily molded; opaque, but wide color range; wide variety of forms——sheets, rods, tubes and laminates.

Epoxides (EP): good toughness, elasticity, chemical resistance and dimensional stability; used as coatings, cements and 'potting' materials for electrical components; easily compounded to cure at room temperatures; widely used in tooling applications.

Acrylonitrile Butadiene Styrene (ABS): contain acrylonitrile, butadiene and styrene; low weight, good strength and very tough; good under severe service conditions.

Polyethylenes (PE): tough; high electrical resistance; used for bottle caps, unbreakable kitchenware and electrical wire insulation.

Polystyrenes (PS): high dimensional stability and low water absorption; best all-around dielectric; burns readily and is adversely affected by citrus juices and cleaning fluids.

Polyamide (nylon) (PA): good abrasion resistance and toughness; excellent dimensional stability; used as bearings with little lubrication; available as monofilaments for textiles, fishing lines, ropes, and so on.

Typical Two-plate Injection Mold

Figure 3-6-4 shows an exploded view of the elements of a typical two-plate injection mold. A brief description of each element is as follows:

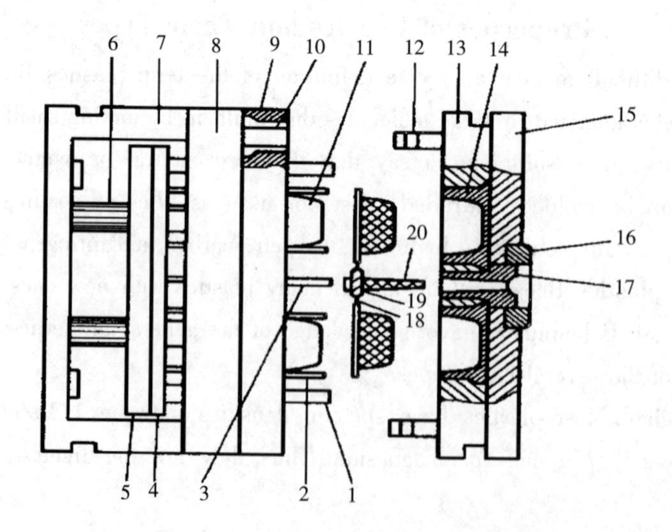

Figure 3-6-4 Conventional two-plate injection mold

(1) Return pin—returns the ejector plate back to its original position when the mold is closed.

(2) Ejector pin—pushes part off the core or out of the cavity.

(3) Sprue puller pin—pulls sprue out of the bushing when mold opens by means of an undercut (not shown).

(4) Ejector retainer plate—plate in which the ejector pins are held.

(5) Ejector plate—often referred to as the ejector cover plate, provides backup for pins set into the ejector-retaining plate.

(6) Support pillar—gives strength and rigidity for the ejector plates.

(7) Ejector housing—provides travel space for the ejector plate and ejector pins.

(8) Support plate—adds rigidity and strength to the plate stackup (not used in all molds).

(9) Core retainer plate—holds the core element, the mating half of the cavity.

(10) Leader pin bushing—provides close tolerance guide for leader pin.

(11) Core—the male part of the cavity.

(12) Leader pin—provides alignment for the two halves of the mold as it opens and closes.

(13) Cavity retainer plate—holds the cavity inserts (the cavity is the area where the part is formed).

(14) Cavity—place where the plastic is formed. Some molds may have the cavity cut directly in the cavity plate rather than by the use of an insert as shown.

(15) Clamp plate—secures the stationary side of the mold to the molding machine. Clamps are inserted in the recess between the clamp plate and the cavity retainer plate.

(16) Locating ring—aligns the nozzle of the injection molding machine with the mold.

(17) Sprue bushing—a conical channel that carries the injected plastic through the top clamp plate to the part or runner system.

(18) Gate—the restricted area of the runner right before the material enters the cavity.

(19) Runner—a passageway for the plastic to flow from the sprue to the part.

(20) Sprue—a passageway for the plastic from the nozzle to the runner.

Chapter 4 Cutting Tool, Fixture and Location

Unit 1 Cutting Tool Design

Text

Physics of metal-cutting provides the theoretical *framework* by which we must examine all other elements of cutting tool design. We have workpiece materials from a very soft, *buttery* consistency to very hard and shear resistant. Each of the workpiece materials must be *handled* by itself; the amount of broad information that is applicable to each workpiece material is reduced as the *distinctions* between workpiece characteristics increase. Not only is there a vast *diversity* of workpiece materials, but there is also a variety of shapes of tools and tool compositions.

The tool designer must *match* the many variables to provide the best possible cutting *geometry*. There was a day when *trial* and error was normal for this decision, but today, with the ever increasing variety of tools, trial and error is far too expensive.

The designer must develop expertise in applying data and making *comparisons* on the basis of the experience of others. For example, tool manufacturers and material salesmen will have figures when their companies have developed. The figures are meant to be guidelines; however, a careful examination of the *literature* available will provide an excellent place from which to start, and be much cheaper than trial and error.

Material removal by machining involves *interaction* of five elements: the cutting tool, the toolholding and guiding device, the workholder, the workpiece, and the machine. The cutting tool may have a single cutting edge or may have many cutting edges. It may be designed for linear or rotary motion. The geometry of the cutting tool depends upon its intended function. The toolholding device may or may not be used for guiding or locating. Toolholder selection is *governed* by tool design and intended function.

The physical composition of the workpiece greatly influences the selection of the machining method, the tool composition and geometry, and the rate of material removal①. The intended shape of the workpiece influences the selection of the machining method and the choice of linear or rotary tool travel. The composition and geometry of the workpiece to a great extent determine the workholder requirements. Workholder selection also depends upon forces produced by the tool on the workpiece. Tool guidance may be incorporated into the workholding function.

Successful design of tools for the material removal processes requires, above all, a complete understanding of cutting tool function and geometry. This knowledge will enable the designer to specify the correct tool for a given task. The tool, in turn, will govern the selection of toolholding and guidance methods. Tool forces govern selection of the workholding device. Although the process involves interaction of the five elements, everything begins with and is based on what happens at the point of contact between the workpiece and the cutting tool.

The primary method of imparting form and dimension to a workpiece is the removal of material by the use of edged cutting tools. An oversize mass is literally carved to its intended shape. The removal of material from a workpiece is termed generation of form by machining, or simply machining.

Form and dimension may also be achieved by a number of alternative processes such as hot or cold extrusion, sand casting, die casting and precision casting. Sheet metal can be formed or drawn by the application of pressure. In addition to machining, metal removal can be accomplished by chemical or electrical methods. A great variety of workpieces may be produced without resorting to a machining operation. Economic considerations, however, usually dictate form generation by machining, either as the complete process or in conjunction with another process.

Cutting tools are designed with sharp edges to minimize rubbing contact between the tool and workpiece. Variations in the shape of the cutting tool influence tool life, surface finish of the workpiece, and the amount of force required to shear a chip from the parent metal. The various angles on a tool compose what is often termed the tool geometry. The tool *signature* or *nomenclature* is a sequence of *alpha* and *numeric* characters representing the various angles, *significant* dimensions, special features, and the size of the *nose* radius[2]. This method of *identification* has been standardized by the American National Standards Institute for carbide and for high speed steel, and is illustrated in Figure 4-1-1, together with the elements that make up the tool signature.

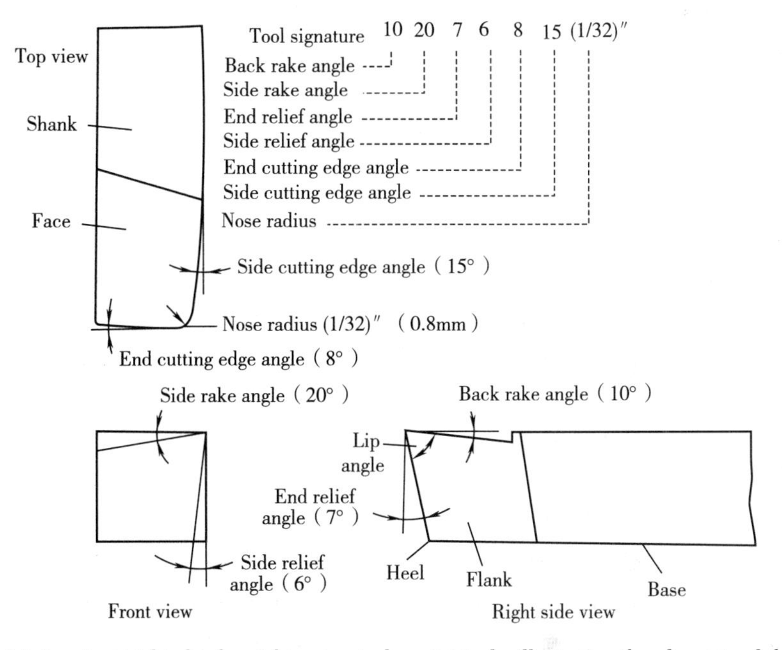

Figure 4-1-1 A straight-shank, right-cut, single-point tool, illustrating the elements of the tool signature as designated by the ANSI. Positive rake angles are shown.

Questions

1. What are the five elements interacting in the machining process?
2. What factor will greatly influence the selection of the cutting tool material and geometry?
3. What factors are influenced by the shape of a cutting tool?

New Words and Expressions

1. framework [ˈfreimwəːk] n. 骨架，结构，组织
2. buttery [ˈbʌtəri] adj. 黄油状的
3. handle [ˈhændl] n. 柄，把手 vt. 管理，处理 vi. 易于操纵
4. distinction [disˈtiŋkʃən] n. 差别，特征，特性
5. diversity [daiˈvəːsiti] n. 参差，多样性，发散
6. match [mætʃ] n. 比赛，匹配 vi. 搭配，和……相称
7. geometry [dʒiˈɔmitri] n. 几何学，几何形状
8. trial [ˈtraiəl] n. 试验，尝试 adj. 试制的
9. comparison [kəmˈpærisn] n. 比较，比喻
10. literature [ˈlitəritʃə] n. 文学，文献，文学作品
11. interaction [ˌintərˈækʃən] n. 相互作用，相互制约
12. govern [ˈgʌvən] vt. 管理，决定 vi. 管理
13. signature [ˈsignitʃə] n. 签名；用法说明
14. nomenclature [nəuˈmenklətʃə] n. 名称，术语，专门用语
15. alpha [ˈælfə] n. 希腊字母的第一个字母，最初，开始
16. numeric [njuːˈmerik] adj. 数值的 n. 数，分数
17. significant [sigˈnifikənt] adj. 有意义的，重要的，有效的
18. nose [nəuz] n. 机头，喷嘴，突出部分
19. identification [aiˌdentifiˈkeiʃən] n. 辨别，鉴别
20. trial and error 反复试验
21. theoretical framework 理论框架
22. on the basis of 在……基础上，根据，主要成分
23. parent metal 母体金属

Notes

[1] The physical composition of the workpiece greatly influences the selection of the machining method, the tool composition and geometry, and the rate of material removal.

工件的物理组成极大地影响着加工方法、刀具材料与几何形状，以及切削速度的选择。

此句两处用到 composition 这个词，其意思是指工件和刀具的“组成成分”。

[2] The tool signature or nomenclature is a sequence of alpha and numeric characters representing the various angles, significant dimensions, special features, and the size of the nose radius.

刀具标识或刀具名称是由希腊字母“α”和一些数字符号构成的有序序列，其中数字分别代表着刀具的各个角度、重要尺寸、特殊性质和刀尖半径。

句中由 representing 引导的分词短语做后置定语，用来修饰 numeric characters。

Glossary of Terms

1. cutting part 切削部分
2. cutting power 切削功率
3. cutting tool 切削工具
4. cutting force 切削力
5. cutting edge 切削刃
6. cutting speed 切削速度
7. depth of cut 背吃刀量
8. tool angle 刀尖角
9. tool back rake angle 刀具背前角
10. tool back clearance angle 刀具背后角
11. tool backlash movement (tool retracting) 退刀
12. tool back wedge angle 刀具背楔角
13. tool base clearance angle 刀具基后角
14. tool center point 刀具中心
15. tool cutting edge angle 刀具主偏角
16. tool cutting edge plane 切削平面
17. tool element (dimensions) 刀具要素（尺寸）
18. tool function 刀具功能
19. tool holder 刀夹，刀柄
20. tool post (tool rest, block) 刀架
21. tool geometrical rake angle 刀具几何前角
22. tool normal clearance (rake) angle 刀具法后角（法前角）
23. tool offset 刀具偏置，调整刀具位置
24. tool clearance (rake, wedge) angle 刀具后角（前角，楔角）
25. toolsetting 刀具调整（或安装）
26. tool box (case, chest) 工具箱
27. cutting process 切削工艺
28. hand tool 手动工具
29. power tool 电动工具

Reading Materials

Chip Formation

The majority of metal-cutting operations involve the separation of small segments or chips from the workpiece to achieve the required shape and size of manufactured parts. Chip formation involves three basic requiremens: ①there must be a cutting tool that is harder and more wear-resistant than the workpiece material; ②there must be interference between the tool and the workpiece as designated by the feed and depth of cut; ③there must be a relative motion or cutting velocity between the tool and the workpiece with sufficient force to overcome the resistance of the workpiece material. As long as these three conditions exist, the portion of the material being machined that interferes with free passage of the tool will be displaced to create a chip.

Many possibilities and combinations exist that may fulfill such requirements. Variations in tool material and tool geometry, feed and depth of cut, cutting velocity, and workpiece material have an effect not only upon the formation of the chip, but also upon the cutting force, cutting horsepower, cutting temperatures, tool wear and tool life, dimensional stability, and the quality of the newly created surface. The interrelationship and the interdependence among these "manipulating factors" constitute the basis for the study of machinability—a study which has been popularly defined as the response of a material to machining.

Figure 4-1-2 illustrates the necessary relationship between the cutting tool and the workpiece for chip formation in several common machining processes. Although it is apparent that different general shapes and sizes of chips may be produced by each of the basic processes, all chips regardless of process are usually classified according to their general behavior during formation.

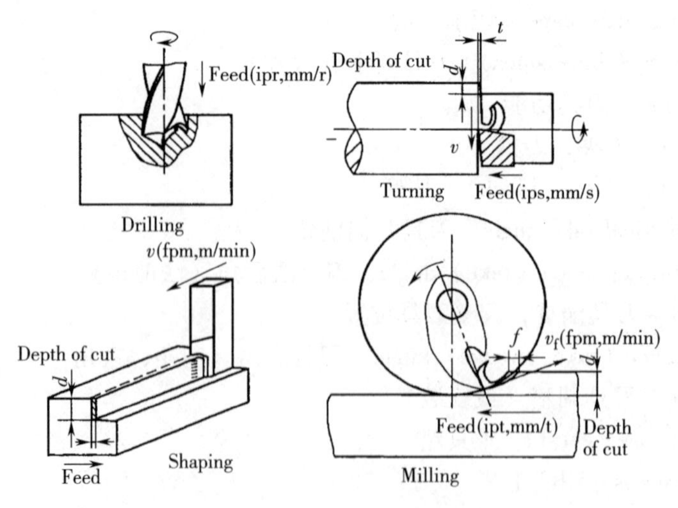

Figure 4-1-2 Examples of feed depth and velocity relationships for several chip-formation processes

Tool Life

The types and mechanisms of tool failure have been previously described. It was shown that excessive cutting speeds cause a rapid failure of the cutting edge; thus, the tool can be declared to have a short life. The rate of tool wear depends on tool and workpiece materials, tool geometry, process parameters, cutting fluids, and the characteristics of the machine tool. Tool wear and the changes in tool geometry during cutting manifest themselves in different ways, generally classified as flask wear, crater wear, nose wear, notching, plastic deformation of the tool tip, chipping, and gross fracture. Other criteria are sometimes used to evaluate tool life:

(1) Change of quality of the machined surface.

(2) Change in the magnitude of the cutting force resulting in changes in machine and workpiece defections causing workpiece dimensions to change.

(3) Change in the cutting temperature.

(4) Costs, including labor costs, tool costs, tool changing time (cost), etc.

The selection of the correct cutting speed has an important bearing on the economics of all metal cutting operations. Fortunately, the correct cutting speed can be estimated with reasonable accuracy from tool life graphs or from the Taylor Tool Life Relationship, provided that necessary data are obtainable.

Mechanics of Cutting

Let's now identify the major independent variables in the cutting process as follows: ①tool material and coatings; ②tool shape, surface finish and sharpness; ③workpiece material and condition; ④cutting speed, feed and depth of cut; ⑤cutting fluids; ⑥characteristics of the machine tool; and ⑦workholding and fixturing.

Dependent variables in cutting are those that are influenced by changes in the independent variables listed above, and include: ①type of chip produced; ②force and energy dissipated during cutting; ③temperature rise in the workpiece, the tool and the chip; ④tool wear and failure, and ⑤ surface finish and surface integrity of the workpiece.

When machining operations yield unacceptable results, normal troubleshooting requires a systematic investigation. A typical question posed is which of the independent variables should be changed first, and to what extent, if ①the surface finish of the workpiece being cut is poor and unacceptable, ②the cutting tool wears rapidly and becomes dull, ③the workpiece becomes very hot, and ④the tool begins to vibrate and chatter.

General recommendations for turning operations are showed in Table 4-1-1.

Table 4-1-1 General recommendations for turning operations

Number	Workpiece material	Cutting tool
1	Low carbon and free machining steels	Uncoated carbide, ceramic-coated carbide, triple-coated carbide, TiN-coated carbide, Al_2O_3 ceramic
2	Medium and high carbon steels	Uncoated carbide, ceramic-coated carbide, triple-coated carbide, TiN-coated carbide, Al_2O_3 ceramic

(Continued)

Number	Workpiece material	Cutting tool
3	Cast iron, gray	Uncoated carbide, ceramic-coated carbide, TiN-coated carbide, Al_2O_3 ceramic
4	Stainless steel, austenitic	Triple-coated carbide, TiN-coated carbide, cermet
5	High-temperature alloys, nickel based	TiN-coated carbide, uncoated carbide, ceramic-coated carbide, Al_2O_3 ceramic, polycrystalline
6	Titanium alloys	Uncoated carbide, TiN-coated carbide
7	Aluminum alloys free machining	Uncoated carbide, TiN-coated carbide, polycrystalline, diamond, cermet
8	Copper alloys	Uncoated carbide, ceramic-coated carbide, triple-coated carbide, TiN-coated carbide, polycrystalline, diamond, cermet
9	Thermoplastics and thermosets	TiN-coated carbide, polycrystalline, diamond

Unit 2　Workholding Principles

Text

The term workholder includes all devices that hold, *grip*, or *chuck* a workpiece to perform a manufacturing operation. The holding force may be applied mechanically, electrically, hydraulically, or *pneumatically*. This section considers workholders used in material-removing operations. Workholding is one of the most important elements of machining processes.

Figure 4-2-1 illustrates almost all the basic elements that are present in a material-removing operation intended to shape a workpiece. The right hand is the toolholder, the left hand is the workholder, the knife is the cutting tool, and the piece of wood is the workpiece. Both hands combine their motions to shape the piece of wood by removing material in the form of chips. The body of the person whose hands are shown may be considered a machine that imparts power, motion, position and control to the elements shown. Except for the element of force multiplication, these basic elements may be found in all of the forms of manufacturing setups where toolholders and workholders are used.

Figure 4-2-2 shows a pair of *pliers* or *tongs* used to hold a rod on which a point has to be ground or filed. This simple workholder illustrates the element of force multiplication by a lever action, and also shows *serrations* on the parts contacting the rod to increase resistance against *slippage*.

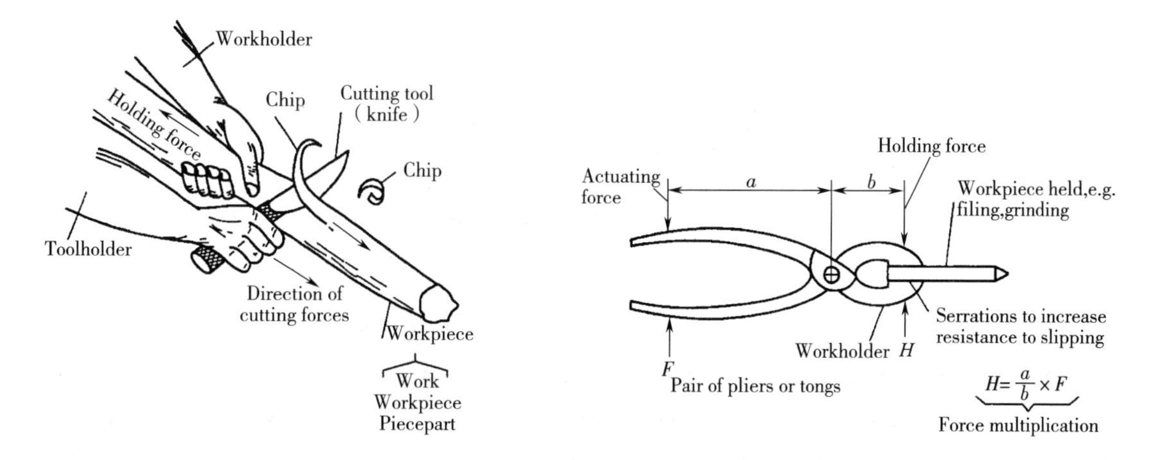

Figure 4-2-1　Principles of workholders

Figure 4-2-2　Multiplication of holding force

Figure 4-2-3 shows a widely used workholder, the screw-operated *vise*. The screw pushes the movable *jaw* and multiplies the applied force. The vise remains locked by the self-locking characteristic of the screw, provides means of *attachment* to a machine, and permits precise placement of the work①.

A vise with a number of refinements often used in workholders is *depicted* in Figure 4-2-4. The main holding force is supplied by hydraulic power, the screw being used only to bring the jaws in contact with a workpiece. The jaws may be replaceable inserts profiled to locate and fit a specific workpiece as shown, and more complicated jaw forms are used to match more complicated workpieces.

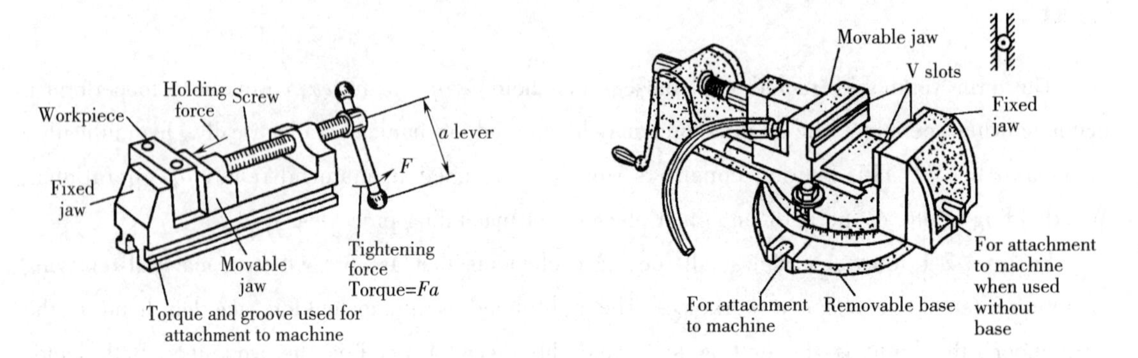

Figure 4-2-3 Elementary workholder (vise)

Figure 4-2-4 Vise with hydraulic clamping

Another large group of workholders are the chucks. They are attached to a variety of machine tools and are used to hold a workpiece during turning, boring, drilling, grinding, and other rotary operations. Many types of chucks are available. Some are tightened manually with a *wrench*, others are power operated by air or hydraulic means or by electric motors. On some chucks, each jaw is individually advanced and tightened, while others have all jaws advance in *unison.* Figure 4-2-5 shows a workpiece clamped in a four-jaw independent chuck. The drill, which is removing material from the workpiece is clamped in a universal chuck.

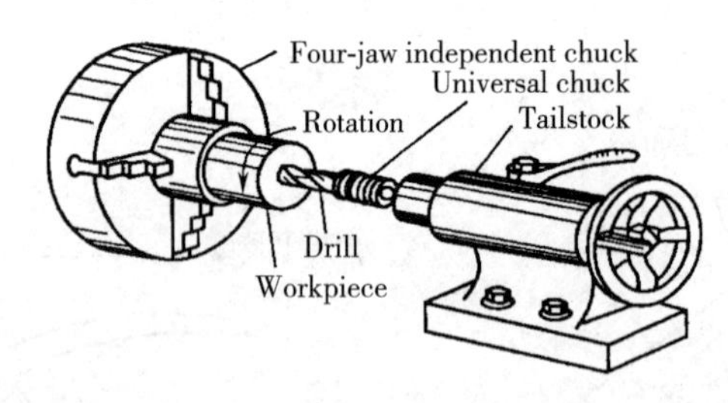

Figure 4-2-5 Holding (chucking) a round workpiece

Purpose and function of workholders. A workholder must position or locate a workpiece in a definite relation to the cutting tool and must withstand holding and cutting forces while maintaining that precise location. A workholder is made up of several elements, each performing a certain function. The locating elements position the workpiece; the structure, or tool body, withstands the forces; the brackets attach the workholder to the machine; the clamps, screws, and jaws apply holding forces. Elements may have manual or power activation. All functions must be performed with the required *firmness* of holding, accuracy of positioning, and with a high degree of safety for the operator and the equipment.

The design or selection of a workholder is governed by many factors, the first being the physical characteristics of the workpiece. The workholder must be strong enough to support the workpiece without deflection. The workholder material must be carefully selected with the workpiece in mind so that neither will be damaged by *abrupt* contact, e. g. , damage to a soft copper workpiece by hard steel jaws②.

Cutting forces imposed by machining operations vary in magnitude and direction. A drilling operation induces *torque*, while a shaping operation causes straight-line *thrust*. The workholder must support the workpiece in opposition to the cutting forces and will generally be designed for a specific machining operation.

Many workholders used in industry are not used on material removing operations. Workholders may be used for the inspection of workpieces, assembly, welding, and so on. There may be very little difference in their basic design and their appearance. Quite often a standard commercial design may be used in one application for a turning operation and for the same or another workpiece in an inspection operation.

Questions

1. What is the purpose and function of the workholder?
2. Describe the principles of workholders.
3. What are the influence factors in design or selection of a workholder?

New Words and Expressions

1. grip [grip] n. 紧握，啮合 vt. 控制
2. chuck [tʃʌk] n. 夹头，夹盘，卡盘 vi. 夹紧，卡紧
3. pneumatic [nju (ː) ˈmætik] adj. 空气的，气动的 n. 气胎
4. pliers [ˈplaiəz] n. 钳，手钳，台虎钳
5. tong [tɔŋ] n. （常用 tongs）钳，夹子 vt. & vi. 用钳夹住
6. serration [seˈreiʃən] n. 锯齿状，成锯齿形
7. slippage [ˈslipidʒ] n. 滑动，下降
8. vise [vais] n. （台）虎钳 vt. 钳住，夹紧
9. jaw [dʒɔː] n. 台虎钳牙，夹片，口部
10. attachment [əˈtætʃmənt] n. 连接物，附加装置
11. depict [diˈpikt] vt. 描写，描述，描绘
12. wrench [rentʃ] n. 拧，扳钳，扳手 vt. 扳紧
13. unison [ˈjuːnizn, -sn] n. 一致，统一
14. firmness [ˈfəːmnis] n. 坚固，坚定，稳固
15. abrupt [əˈbrʌpt] adj. 突然的，陡峭的
16. torque [tɔːk] n. 转矩，扭矩
17. thrust [θrʌst] vt. 猛推，延伸 n. 拉力，牵引力

18. except for 除了，只有
19. in contact with 和……接触着
20. tool-holder 刀夹，刀杆，刀柄

Notes

[1] The vise remains locked by the self-locking characteristic of the screw, provides means of attachment to a machine, and permits precise placement of the work.

台虎钳是依靠（丝杠）螺纹的自锁特性来保持锁紧的，它还可以连同其他部件一起附着在机床上，并确保加工时的精确定位。

句中 remains，provides，permits 为并列谓语，逐步说明台虎钳的作用。

[2] The workholder material must be carefully selected with the workpiece in mind so that neither will be damaged by abrupt contact, e. g. , damage to a soft copper workpiece by hard steel jaws.

在考虑工件材料的前提下，必须谨慎选择夹具材料，以防止二者产生突发接触性破坏。例如：若台虎钳口选用硬的钢质材料，就会使较软的铜质工件受到破坏。

句中 so that neither will be damaged by abrupt contact 为目的状语从句。

Glossary of Terms

1. workpiece 工件，冲压件
2. holder of punch 凸模夹持器
3. locating device 定位装置
4. locating face 定位面
5. locating pin 定位销（挡料销）
6. locating plate 定位板
7. locating ring 定位圈
8. locating ruler 定位尺
9. locating element 定位零件（定位要素）
10. locating principle 定位原理
11. workholding 工件夹紧
12. work hardening 加工硬化
13. internal force 内力
14. clamping surface 固定面
15. hole scraping (turning, milling, lapping) 刮孔（车孔、铣孔、研孔）
16. hole grinding (slotting, honing, flanging) 磨孔（插孔、珩孔、翻孔）
17. grip device (clamping device) 夹紧装置
18. grip holder 夹头
19. flat surface 平面
20. cylindrical surface 圆柱面

21. irregular surface 不规则面
22. plane location 平面定位
23. concentric location 定心定位
24. radial location 径向定位
25. face chuck 平面夹盘
26. universal chuck 万能夹头

Reading Materials

Locating Principles

To insure successful operation of a workholding device, the workpiece must be accurately located to establish a definite relationship between the cutting tool and some points or surfaces of the workpiece. This relationship is established by locators in the workholding device which position and restrict the workpiece to prevent its moving from its predetermined location. The workholding device will then present the workpiece to the cutting tool in the required relationship. The locating device should be so designed that each successive workpiece, when loaded and clamped, will occupy the same position in the workholding device. Various methods have been devised to effectively restrict the movement of workpieces. The locating design selected for a given workholding device will depend on the nature of the workpiece, the requirements of the metal-removing operation to be performed, and other restrictions on the workholding device.

Types of Location

Basic workpiece location can be divided into three fundamental categories: plane, concentric and radial. In many cases, more than one category of location may be used to locate a particular workpiece. However, for the purpose of identification and explanation, each will be discussed individually.

Plane location is normally considered the act, or process, of locating a flat surface. But many times irregular surfaces may also be located in this manner. Plane location is simply locating a workpiece with reference to a particular surface or plane (Figure 4-2-6).

Concentric location is the process of locating a workpiece from an internal or external diameter (Figure 4-2-7).

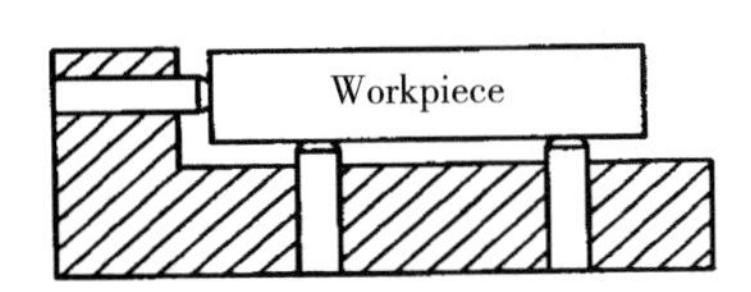

Figure 4-2-6 Plane location

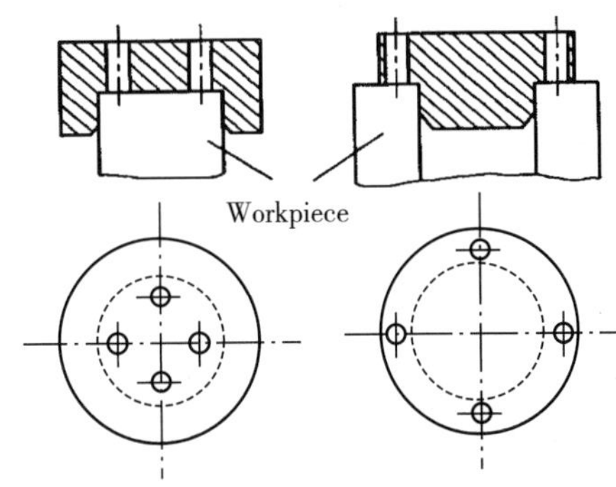

Figure 4-2-7 Concentric location

Radial location is normally a supplement to concentric location. With radial location (Figure 4-2-8), the workpiece is first located concentrically and then a specific point on the workpiece is located to provide a specific fixed relationship to the concentric locator.

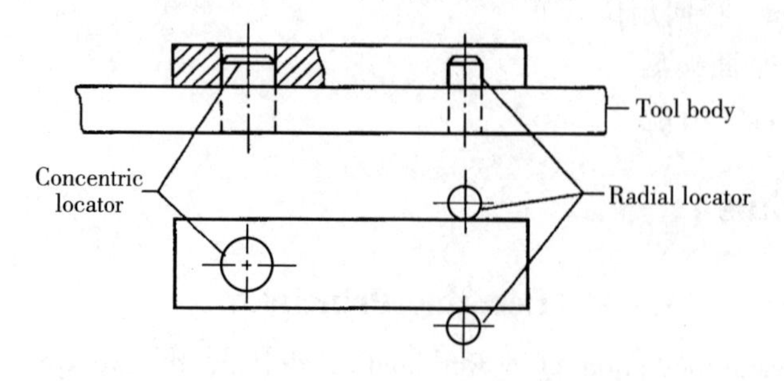

Figure 4-2-8 Radial location

Most workholders use a combination of locational methods to completely locate a workpiece. The part shown in Figure 4-2-9 is an example of all three basic types of location being used to reference a workpiece.

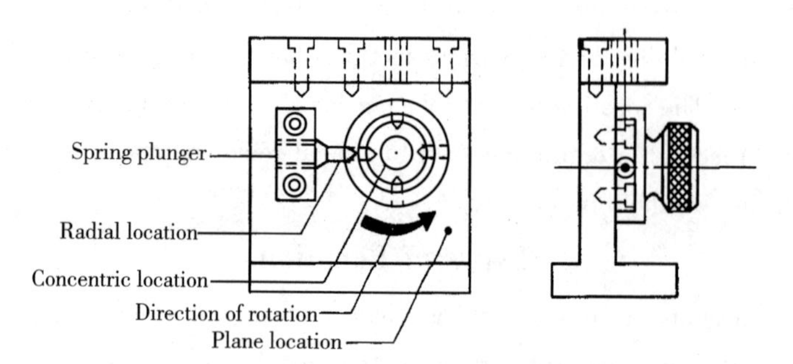

Figure 4-2-9 Plane, concentric and radial location

Unit 3 Jig and Fixture Design

Text

Jigs are workholders which are designed to *hold*, *locate* and support a workpiece while guiding the cutting tool throughout its cutting cycle. Jigs can be divided into two general classifications: drill jigs and boring jigs. Of these, drill jigs are, by far, the most common. Drill jigs are generally used for drilling, *tapping* and *reaming*, but may also be used for *countersinking*, *counterboring*, *chamfering*, and spotfacing. Boring jigs, on the other hand, are normally used exclusively for boring holes to a *precise*, *predetermined* size. The basic design of both classes of jigs is essentially the same. The only major difference is that boring jigs are normally fitted with a *pilot bushing* or bearing to support the outer end of the boring bar during the machining operation.

In designing any jig, there are numerous considerations that must be addressed. Although several of these points, such as locating, supporting and clamping, have already been covered, they are included in this section because they apply to jig design. Since all jigs have a similar construction, the points covered for one type of jig normally apply to the other types as well. Jig design and selection begins with an analysis of the workpiece and the manufacturing operation to be performed.

One of the first considerations in the design of any workholder is the relative balance between the cost of the tool and the expected *benefits* of using the tool for production①. All workholders should save more in production costs than the tool costs to design and construct. In many instances, tool designers may have to complete detailed estimates to justify the cost of special tooling. This involves a close look at the part drawing, process specifications and other related documents.

Typically, the complexity of the part, location and number of holes, required accuracy, and the number of parts to be made are all points which must be considered to determine if the cost of a particular jig is *warranted*. Once the tool designer is satisfied that the cost of special tooling is justified, the remaining data required to produce a suitable workholder is *compiled* and analyzed.

Fixtures are workholders which are designed to hold, locate, and support the workpiece during the machining cycle. Unlike jigs, fixtures do not guide the cutting tool, but rather provide a means to reference and *align* the cutting tool to the workpiece. Fixtures are normally classified by the machine with which they are designed to be used. A sub-classification is sometimes added to further specify the fixture classification, this sub-classification identifies the specific type of machining operation the fixture is intended to perform. For example: a fixture used with a milling machine is called a milling fixture, however, if the operation it is to perform is *gang* milling, it may also be called a gang-milling fixture. Likewise, a *bandsawing* fixture designed for *slotting* operations may also be referred to as a bandsaw-slotting fixture.

The similarity between jigs and fixtures normally ends with the design of the tool body. For the most part, fixtures are designed to withstand much greater stresses and tool forces than jigs, and are always securely clamped to the machine②. For these reasons, the designers must always be aware of proper locating, supporting and clamping methods when fixing any part.

In designing any fixture, there are several considerations in addition to the part which must be addressed to complete a successful design. Cost, production capabilities, production processing and tool longevity are some of the points which must *share* attention with the workpiece when a fixture is designed.

As with all tooling, the first consideration in fixture design is the cost *versus* the benefit. The production quantity, rate, or accuracy must warrant the added expense of special tooling. In addition, the fixture must pay for itself with savings derived from its use in as short a time as possible.

Questions

1. What is a jig?
2. What are the functions of a fixture?
3. What is the difference between a jig and a fixture?

New Words and Expressions

1. jig [dʒig] n. 夹具，模具，规尺
2. hold [həuld] vt. & n. 拿，握，夹住
3. locate [ləuˈkeit] vt. 设置，定位
4. tap [tæp] n. 丝锥，塞子
5. reamer [ˈriːmə] n. 扩孔锥，铰刀，铰床
6. countersink [ˈkauntəsiŋk] n. 埋头孔，锥口孔 vt. 打埋头孔于
7. counterbore [kauntəˈbɔː] n. 扩孔，沉孔
8. chamfer [ˈtʃæmfə] n. 斜面，倒角 vt. 去角
9. precise [priˈsais] adj. 正确的，精确的，精密的
10. predetermine [ˌpriːdiˈtəːmin] vt. 预定
11. pilot [ˈpailət] n. 领航员，定位销
12. bush [buʃ] n. 衬套，轴瓦 vt. 加衬套于……
13. benefit [ˈbenifit] n. 利益，好处 vt. 有益于…… vi. 受益
14. warrant [ˈwɔrənt] vt. 证明，理由，根据；保证，批准
15. compile [kəmˈpail] vt. 编辑，编译程序
16. fixture [ˌfikstʃə] n. 夹具，夹紧装置
17. align [əˈlain] vt. 匹配，排列成一行，定位
18. gang [gæŋ] n. 一组，一队 v. 成群结队
19. bandsaw [ˈbændsɔː] n. 带锯，带锯机 vt. 用带锯锯
20. slotter [ˈslɔtə] n. 插床，立刨床

21. share [ʃɛə] n. 一份，股份 vt. 均分，分配
22. versus [ˈvəːsəs] prep. 与……比较，以……为转移
23. be aware of 知道，意识到，认识

Notes

[1] One of the first considerations in the design of any workholder is the relative balance between the cost of the tool and the expected benefits of using the tool for production.

在设计工件夹持装置时，首先要考虑的问题之一是制造该夹具的成本和使用该夹具进行生产所希望产生的效益两者之间应保持相对平衡。

单个分词 expected 做名词 benefits 的前置定语；动名词短语 using the tool for production 做介词 of 的宾语。

[2] For the most part, fixtures are designed to withstand much greater stresses and tool forces than jigs, and are always securely clamped to the machine.

在绝大部分情况下，fixtures（夹紧装置）设计得比 jigs（夹具）能够承受大得多的应力和切削力，并且总能牢固地固定在机床上。

介词短语 for the most part 在这里相当于 in most cases，可译为“在绝大部分情况下”。

Glossary of Terms

1. spot face 孔口平面
2. drill and countersink 中心钻
3. counterbore cutter head 扩孔钻头
4. jig boring machine 坐标镗床
5. jig grinding machine 坐标磨床
6. jig drill 钻模钻床
7. jig and fixture 钻模与夹具
8. fixture of gear cutting machine 齿轮加工机床夹具
9. fixture of grinding machine 磨床夹具
10. fixture of milling machine 铣床夹具
11. fixture of planning machine 刨床夹具
12. fixture of slotting machine 插床夹具
13. vacuum fixture 真空夹具
14. universal fixture (jig) 通用夹具
15. stationary fixture 固定夹具
16. standard fixture (jig) 标准夹具
17. pneumatic fixture (jig) 气动夹具
18. open-side boring and milling machine 悬臂镗铣床
19. magnetic fixture (jig) 磁力夹具
20. hydraulic fixture (jig) 液压夹具

21. flexible fixture 柔性夹具
22. modular fixture 组合夹具
23. fixture assembly 夹具组合件
24. general purpose machine tool (GPM) 通用机床
25. special purpose machine tool (SPM) 专用机床
26. sawing machine 锯床
27. jaw chuck 爪式卡盘
28. side tool head 侧刀架
29. turning head 多刀转塔
30. turret head 六角(转塔)刀架

Reading Materials

Types of Jigs

Jigs are made in a wide variety of different styles. The specific style or type of jig which should be used for a particular application is generally determined by the workpiece itself. The two general categories of jigs are open and closed. Open jigs are normally used for parts which only require machining on a single surface, or side. Closed jigs are used to machine parts which require operations on more than one side or surface. The terms commonly used to describe specific types of jigs within each category are generally associated with the basic appearance or construction of the jig.

Template jigs. Template jigs are generally flat, open jigs which are used to locate the general position of drilled holes. This type of jig does not normally contain any device for clamping or securing the jig to the workpiece, but relies on auxiliary clamps to hold the jig when necessary. Template jigs are the least expensive type of jig, and are frequently used where extreme accuracy is not required. When used for limited production, template jigs do not normally contain drill bushings. Instead, the entire jig plate is hardened. Figure 4-3-1 shows several examples of applications for which template jigs are well suited.

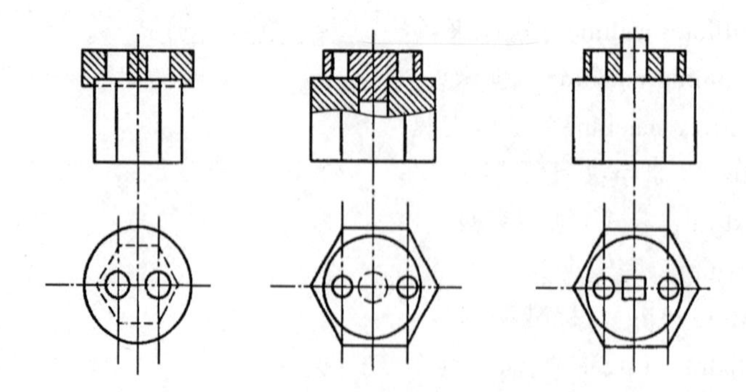

Figure 4-3-1 Applications for template jigs

Plate jigs. Plate jigs are a somewhat more sophisticated variation of the basic template jig. These jigs are generally more accurate and durable than template jigs and also include a means to securely clamp the workpiece. The principal variations of the plate jig commonly used for stationary workholders are the plain plate jig, table jig and the sandwich jig.

Plain plate jigs. Plain plate jigs (Figure 4-3-2) consist of a jig plate which forms the main body of the jig and contains the drill bushings (when used), locators and clamping device. Depending on the complexity of the workpiece, it is often less expensive to make several plate jigs than a single box, or closed jig.

Table jigs. Table jigs (Figure 4-3-3) are another variation of the plate jig which uses the jig plate as a main member with the other components attached. The principal difference is the addition of the legs which raise the jig off the machine table.

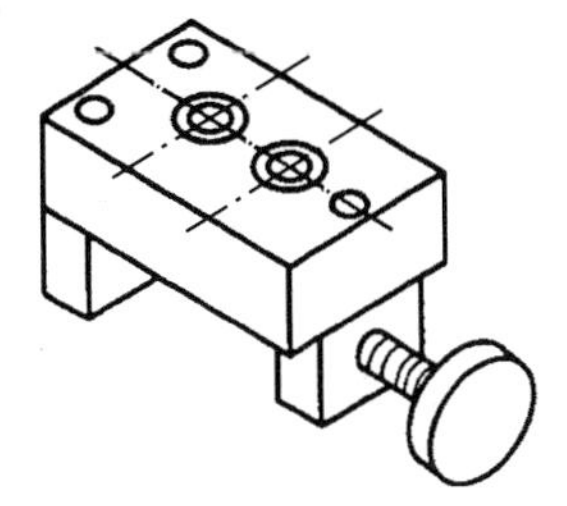

Figure 4-3-2 Plain plate jig

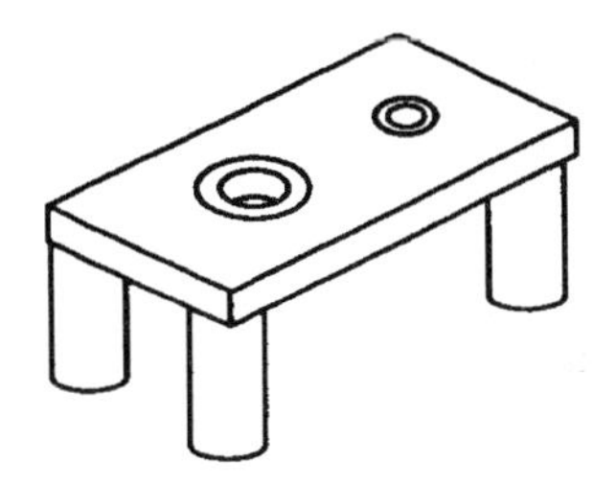

Figure 4-3-3 Table jig

Sandwich jigs. Sandwich jigs (Figure 4-3-4) consist of two plates which are used to sandwich the workpiece. These jigs use a jig plate to establish the location of the drill bushing and the position of the part to be drilled.

Angle plate jigs. Angle plate jigs (Figure 4-3-5) are mainly used to drill or machine parts at an angle to the locating or reference surface. The two principal variations of the angle plate design are the plain angle plate jig and the modified angle plate jig. The plain angle plate jig is intended to machine parts at right angles to the mounting axis, while the modified angle plate jig is designed to accommodate angles other than 90°.

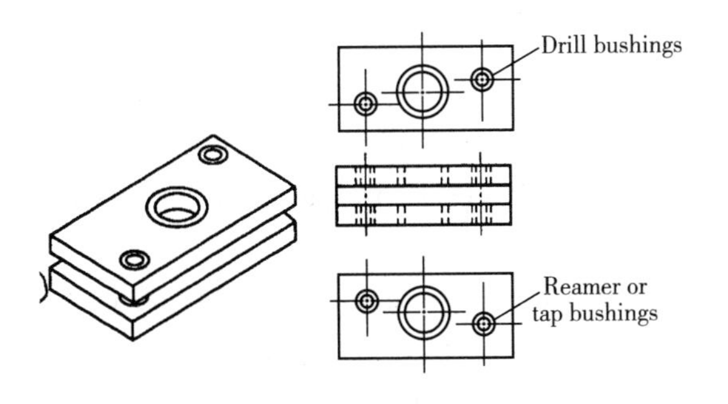

Figure 4-3-4 Sandwich jig

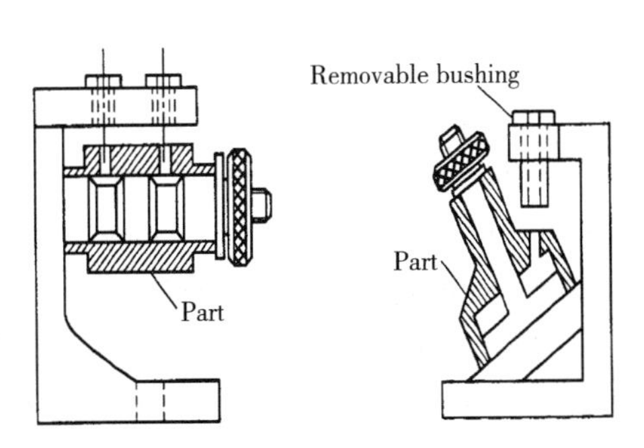

Figure 4-3-5 Angle plate jig

Leaf jigs. Leaf jigs are similar to sandwich jigs but have a hinged jig plate rather than a two-piece construction. Depending on their design, this type of jig may use any number of different devices to secure the leaf during the machining cycle. Figure 4-3-6 shows two of the more common methods, using a quarter turn screw and a cam lock lever.

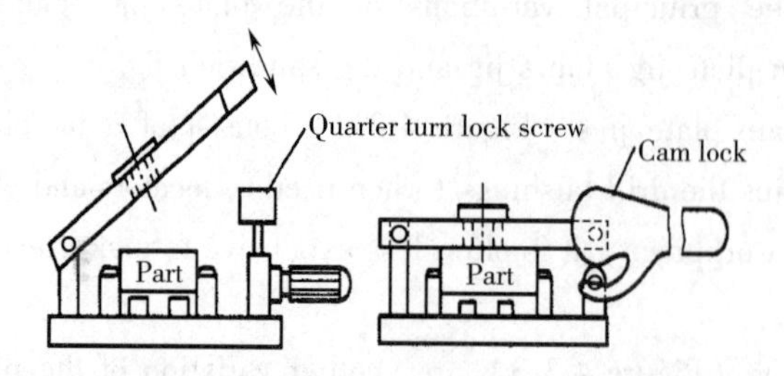

Figure 4-3-6 Two common methods to secure leaf jigs

Box jigs. Box or tumble jigs are constructed to completely enclose a workpiece. As shown in Figure 4-3-7, box jigs are capable of machining a part from almost any angle and on every surface. These jigs are the most expensive type of jig, but they greatly reduce the amount of part handling since the part can be machined completely without removing it from the jig.

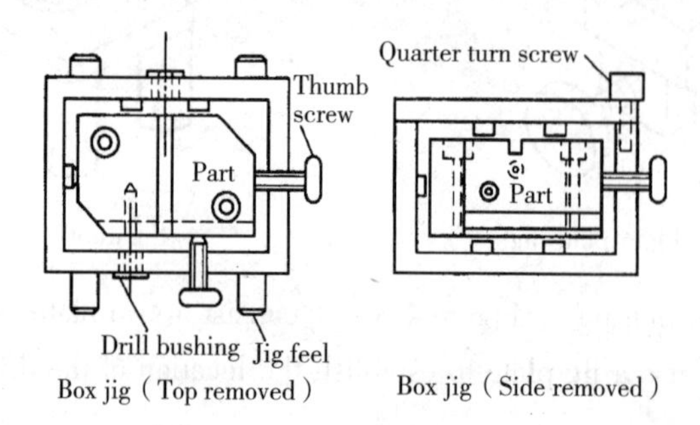

Figure 4-3-7 Box jigs provide for machining from almost any angle

Types of Fixtures

Fixtures are classified either by the machine they are used on, or by the process they perform on a particular machine tool. However, fixtures also may be identified by their basic construction features. For example, a lathe fixture made to turn radii is classified as a lathe radius turning fixture. But if this same fixture were a simple plate with a variety of locators and clamps mounted on a faceplate, it is also a plate fixture. Like jigs, fixtures are made in a variety of different forms.

While many fixtures use a combination of different features, almost all can be divided into five distinct groups. These include plate fixtures, angle plate fixtures, vise jaw fixtures, indexing fixtures, and multi-part, or multi-station fixtures.

Plate fixtures, as their name implies, are constructed from a plate with a variety of locators, supports and clamps (Figure 4-3-8). Plate fixtures are the most common type of fixture. Their versatility makes them adaptable for a wide range of different machine tools. Plate fixtures may be made from any number of different materials, depending on the application of the fixture.

The angle plate fixture (Figure 4-3-9) is a modified form of plate fixture. Here, rather than having a reference surface parallel to the mounting surface, the angle plate fixture has a reference surface perpendicular to its mounting surface. This construction is very useful for those machining operations which are performed perpendicular to the primary reference surface of the fixture.

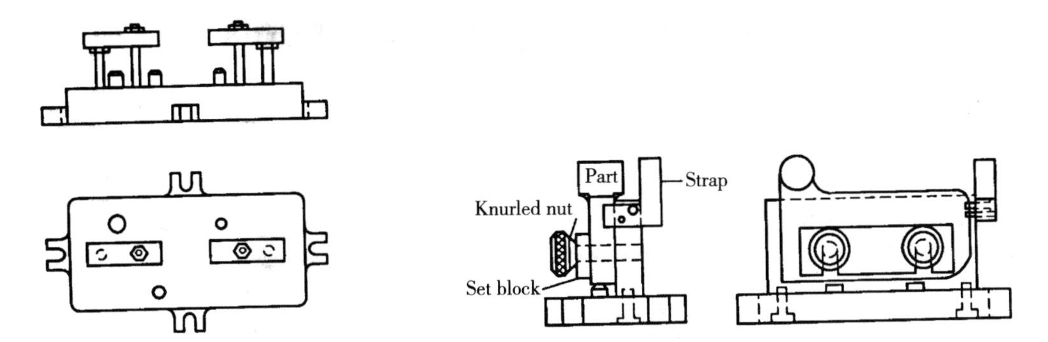

Figure 4-3-8 Plate fixture

Figure 4-3-9 Angle plate fixture

Vise jaw fixtures are basically modified vise jaw inserts which are machined to suit a particular workpiece. In use, these modified vise jaws are installed in place of the standard, hardened jaws normally furnished with milling machine vises. Vise jaw fixtures are the least expensive type of fixture to produce, and since there are so few parts involved, they are also the simplest to modify.

Indexing fixtures, like indexing jigs, are used to reference workpieces which must have machine details located at prescribed spacings.

Chapter 5 Machine Tools

Unit 1 Lathes

Text

The oldest and most common machine tool is the *lathe*, which removes material by rotating the workpiece against a single-point cutter①. Lathes used in manufacturing can be classified as speed lathes, engine lathes, toolroom lathes, turret lathes, automatic lathes, tracer lathes and numerical control turning centers.

Engine lathe. The engine lathe is so named because it was originally powered by the steam "engine" of the eighteenth and nineteenth century where the lathe was connected to *pulley* power from overhead line shafts or connected directly to the steam engine. Today lathes are equipped with electric motors. Although simple and *versatile*, an engine lathe requires a skilled machinist because all controls are *manipulated* by hand. Consequently, it is inefficient for repetitive operations and for large production runs.

The essential components of an engine lathe (Figure 5-1-1) are the bed, *headstock assembly*, *tailstock* assembly, *carriage* assembly, quick-change *gearbox*, and the lead screw and feed rod.

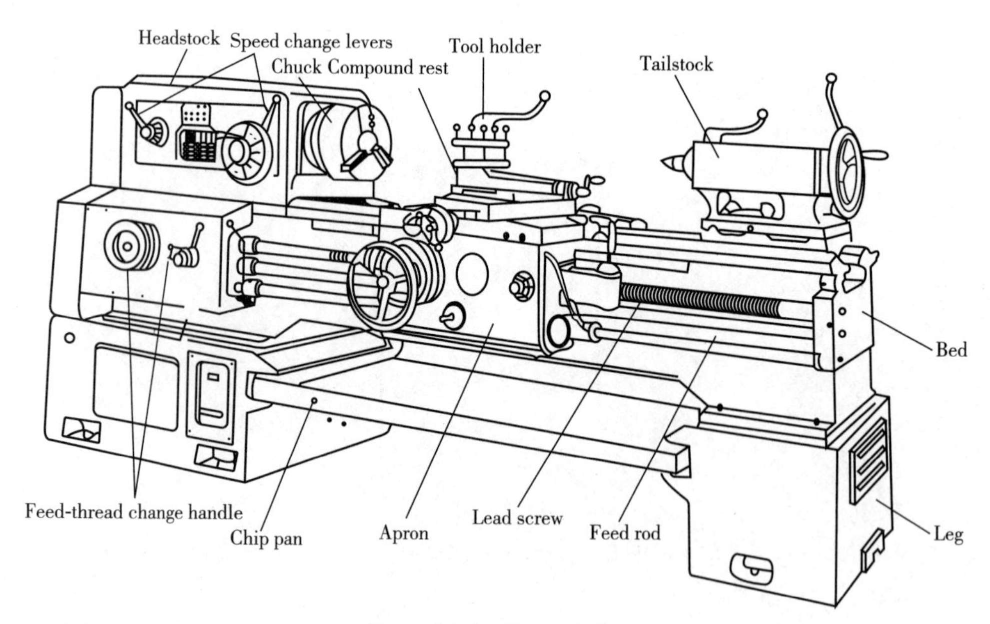

Figure 5-1-1 Engine lathe

The bed supports all the major components of the lathe. Beds have a large mass and are rigidly built, usually from grey or nodular cast iron. The top portion of the bed has two ways, with various cross-sections, which are surface-hardened and precision-machined for wear resistance and dimensional accuracy during use.

The headstock is fixed to the bed and is equipped with motors, pulleys and V-belts that supply power to the spindle at various rotational speeds. The speeds, which are arranged in logical geometric progression, can be set through manually-controlled selectors. Most headstocks are equipped with a set of gears, and some have various drives to provide a continuously variable speed range to the spindle. Headstocks have a hollow spindle to which work holding devices, such as chucks and *collets*, are attached, and long bars or tubing can be fed through for various turning operations[②]. The accuracy of a lathe depends greatly on the spindle. It carries the workholders and is mounted in accurate bearings, usually *preloaded* tapered roller or ball types.

The tailstock can slide along the ways of the lathe to accommodate different lengths of stock and be clamped at any position. It is commonly provided with a center that may be fixed (dead center) or may be free to rotate with the workpiece (live center). Drills and reamers can be mounted on the tailstock *quill* (a hollow cylindrical part with tapered hole) to drill axial holes in the workpiece. The tailstock can be adjusted in and out with respect to the center line of the bed. This allows for adjusting the alignment of the centers and for taper turning.

The lead screw is a long, carefully threaded shaft located slightly below and parallel to the bed ways extending from the headstock to the tailstock. It is geared to the headstock and its rotation may be *reversed*; it is fitted to the carriage assembly and may be engaged or released from the carriage during cutting operations. The lead screw is for cutting threads and *disengaged* when not in use to preserve its accuracy. Below the lead screw is a feed rod, which transmits power from the quick change box to drive the *apron* mechanism for cross and longitudinal power feed. Changing the speed of the lead screw or feed rod is done at the quick-change gearbox located at the headstock end of the lathe.

The carriage assembly includes the compound rest, tool saddle and apron. Since it supports and guides the cutting tool, the carriage assembly must be rigid and constructed with accuracy.

Workholding devices are important in machine tools. In a lathe, one end of the workpiece is clamped to the spindle by a chuck, collet, face plate, or *mandrel* or between centers. A chuck is usually equipped with three or four jaws.

Questions

1. What is an engine lathe?
2. What are essential components of an engine lathe?
3. Name and describe the major units of lathes.
4. What movements does the tool have on a lathe? How are they obtained?
5. From where does the power feed come on a lathe? How is it varied?
6. How is the carriage moved for cutting threads on a lathe? For other operations?

New Words and Expressions

1. lathe [leið] n. 车床 vt. 用车床加工
2. pulley [ˈpuli] n. 滑轮，滑车，带轮 vt. 用滑车推动（举起）
3. versatile [ˈvəːsətail] adj. 多方面的，多用途的，万用的
4. manipulate [məˈnipjuleit] vt. 操纵，处理，应付
5. headstock [ˈhedstɔk] n. 主轴箱，床头箱
6. assembly [əˈsembli] n. 装配，组合，装配图
7. tailstock [ˈteilstɔk] n. 尾座，尾架
8. carriage [ˈkæridʒ] n. 留板，拖板，车架
9. gearbox [ˈgiəbɔks] n. 齿轮箱，变速箱
10. collet [ˈkɔlit] n. 夹头，套筒
11. preload [ˌpriːˈləud] n. & v. 预载，预加负荷
12. quill [kwil] n. 钻轴，衬套 vt. 卷在线轴上
13. reverse [riˈvəːs] vt. 倒转，反转 n. 反向，换向 adj. 相反的
14. disengage [ˌdisinˈgeidʒ] vt. 脱开，脱离，使分离
15. apron [ˈeiprən] n. 挡板，机床拖板箱，溜板箱
16. mandrel [ˈmændril] n. 心轴，心棒
17. be connected with（to） 与……有关系
18. be equipped with 用……装备，设有……装置
19. in reverse 相反，反之
20. precision-machined 精密（机械）加工的
21. be arranged in 把……排列成，按……来排列

Notes

[1] The oldest and most common machine tool is the lathe, which removes material by rotating the workpiece against a single-point cutter.

最古老和最通用的机床是车床。车床通过工件相对于单刃车刀的旋转来实现材料切削。

句中由 which 引导出非限定性定语从句，which 代表前句中的 lathe，并通过从句对其进行补充说明，且 which 在从句中做主语。

[2] Headstocks have a hollow spindle to which work holding devices, such as chucks and collets, are attached, and long bars or tubing can be fed through for various turning operations.

主轴箱有一个空心轴，可用来安装工件夹紧装置，如卡盘和弹簧夹头，并且长棒和长管类零件能够通过空心轴进给而进行各种车削加工。

句中由 which 引导一个限定性定语从句，to which work holding devices, such as chucks and collets, are attached 修饰 hollow spindle；such as 引导同位语，进一步说明 work holding devices。

Glossary of Terms

1. machine tool 机床
2. multi-tool lathe 多刀车床
3. engine lathe 普通车床
4. vertical lathe 立式车床
5. toolroom lathe 工具车床
6. turret lathe 转塔车床
7. tracer-controlled lathe 仿形（或靠模）控制车床
8. automatic lathe 自动车床
9. universal lathe 万能车床
10. lead screw 丝杠，导（螺）杆
11. power chuck 动力卡盘
12. single spindle automatic lathe 单轴自动车床
13. ultraprecision machining 超精密加工
14. lathe control system 车床控制系统
15. lathe transmission system 车床传动系统
16. lathe cutting tool 车床切削刀具
17. lathe carriage 车床溜板
18. lathe center 车床顶尖
19. lathe chuck 车床卡盘
20. lathe spindle 车床主轴
21. lathe accessories (attachment) 车床附件（附具）
22. lathe bed 车床床身
23. lathe dog 鸡心夹头；卡箍
24. three-jaw self-centering chucks 自定心卡盘
25. four-jaw independent chucks 单动卡盘
26. face-plate lathe 落地车床
27. headstock assembly 主轴箱组件
28. tailstock assembly 尾座组件
29. quick-change gearbox 快速变速箱
30. tool post (apron) 刀座，刀架

Reading Materials

Turning Processes

Turning processes are very versatile. The following processes are capable of producing a wide variety of shapes illustrated in Figure 5-1-2:

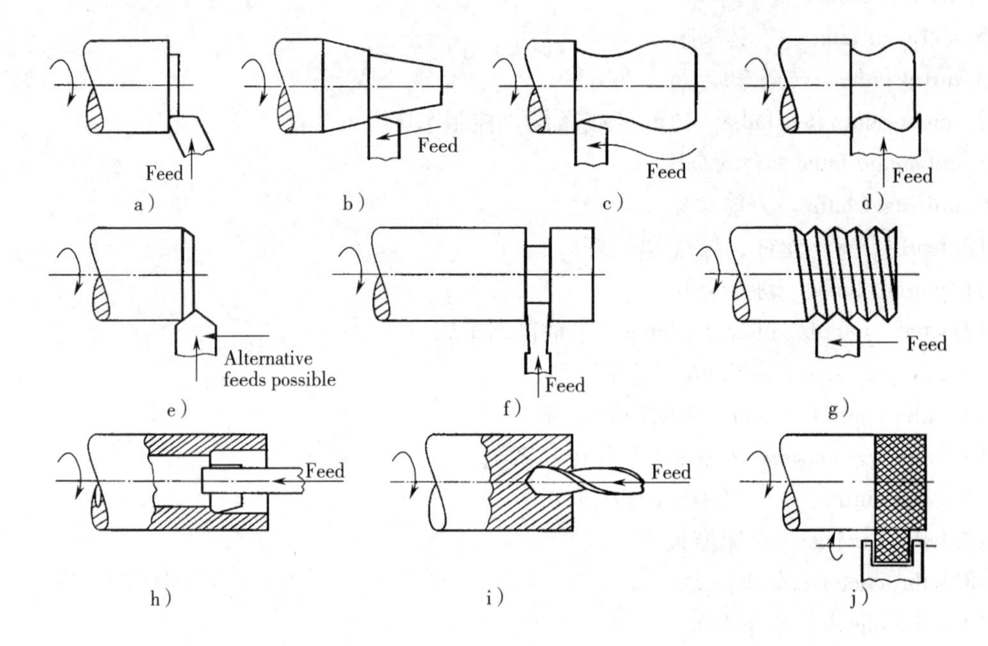

Figure 5-1-2 Machined operations other than turning that are performed on a lathe
a) Facing b) Taper turning c) Contour turning d) Form turning e) Chamfering f) Cutoff
g) Threading h) Boring i) Drilling j) Knurling

(1) Facing—The tool is fed radially into the rotating work on one end to produce a flat surface on the other end.

(2) Taper turning—Instead of feeding the tool parallel to the axis of rotation of the work, the tool is fed at an angle, thus creating a tapered cylinder or conical shape.

(3) Contour turning—Instead of feeding the tool along a straight line parallel to the axis of rotation as in turning, the tool follows a contour that is other than straight, thus creating a contoured form in turned part.

(4) Form turning—In this operation, sometimes called forming, the tool has a shape that is imparted to the work by plunging the tool radially into the work.

(5) Chamfering—The cutting edge of the tool is used to cut an angle on the corner of the cylinder, forming what is called a "chamfer".

(6) Cutoff—The tool is fed radially into the rotating work at some location along its length to cut off the end of the part. This operation is sometimes referred to as parting.

(7) Threading—A pointed tool is fed linearly across the outside surface of the rotating workpart in a direction parallel to the axis of rotation at a large effective feed rate, thus creating threads in the cylinder.

(8) Boring—A single point tool is fed linearly, parallel to the axis of rotation, on the inside diameter of an existing hole in the part.

(9) Drilling—Drilling can be performed on a lathe by feeding the drill into the rotating work along its axis. Reaming can be performed in a similar way.

(10) Knurling—This is not a machining operation because it does not involve cutting of material. Instead, it is a metal forming operation used to produce a regular cross hatched pattern in the work surface.

Turning is performed at various ① rotational speeds, N, of thc workpiecc clampcd in a spindle, ②depths of cut, d, and ③feeds, f, depending on the workpiece materials, cutting-tool materials, surface finish and dimensional accuracy required, and the characteristics of the machine tool.

Lathe Size

Lathe size is expressed in terms of the diameter of the work it will swing. A 400mm lathe has sufficient clearance over the bed rails to handle work 400mm in diameter. A second dimension is necessary to define capacity in terms of workpiece length. Some manufactures use maximum work length between the lathe centers, whereas others express it in terms of bed length. The diameter that can be turned between centers is somewhat less than the swing because of the allowance for the carriage.

The specifications of a CA6140 engine lathe are given as follows:

Swing over bed:	400mm
Swing over cross slid:	210mm
Distance between centers:	750 ~ 2000mm
Number of spindle speeds:	24
Range of spindle speeds :	10 ~ 1400r/min
Number of feeds:	64
Range of longitudinal feeds:	0. 028 ~ 6. 33mm/rev.
Range of cross feeds:	0. 014 ~ 3. 16mm/rev.
Thread range-metric:	1 ~ 192mm
Thread range-inch:	2 ~ 24 tpi
Thread range-diameter:	1 ~ 96 D. P.
Thread range-modular:	0. 25 ~ 48 Mod.
Hole through spindle:	50mm
Spindle taper:	6 Morse
Motor:	7. 5kW

Cutting Fluids

Cutting fluids have been used extensively in machining operation to achieve the following results:

(1) Reduce friction and wear, thus improving tool life and the surface finish of the workpiece.

(2) Cool the cutting zone, thus improving tool life and reducing the temperature and thermal distortion of the workpiece.

(3) Reduce force and energy consumption.

(4) Flush away the chips from the cutting zone, and thus prevent the chips from interfering with the cutting process, particularly in operations such as drilling and tapping.

(5) Protect the machined surface from environmental corrosion.

Depending on the type of machining operation, the cutting fluid needed may be a coolant, a lubricant, or both. The effectiveness of cutting fluids depends on a number of factors, such as the type of machining operation, tool and workpiece material, cutting speed, and the method of application. Water is an excellent coolant and can reduce effectively the high temperatures developed in the cutting zone. However, water is not an effective lubricant, hence it does not reduce friction. Furthermore, it causes the rusting of workpieces and machine-tool components. On the other hand, as we have seen, effective lubrication is an important factor in machining operations.

The need for a cutting fluid depends on the severity of the particular machining operation, which may be defined as the level of temperatures and forces encountered, the tendency for built-up edge formation, the ease with which chips produced can be removed from the cutting zone, and how effectively the fluids can be applied to the proper region at the tool-chip interface. The specific machining processes of relative severity in increasing order of severity are: sawing, turning, milling, drilling, gear cutting, thread cutting, tapping and internal broaching.

Unit 2 Milling Machines

Text

Milling machines must provide a rotating spindle for the cutter and a table for fastening, positioning and feeding the workpiece. Various machine tool designs satisfy these requirements. To begin with, milling machines can be classified as horizontal or vertical. A horizontal milling machine has a horizontal spindle, and this design is well-suited for performing peripheral milling (e. g. , slab milling, slotting, side and straddle milling) on workpieces that are roughly cube shaped. A vertical milling machine has a vertical spindle, and this *orientation* is appropriate for face milling, end milling, surface contouring, and die sinking on relatively flat workpieces.

Other than spindle orientation, milling machines can be classified into four types: *knee*-and-*column*, bed type, planer type, and special type milling machines.

Knee-and-column milling machines. The knee-and-column milling machine is the basic machine tool for milling. This type of machine is capable not only of doing straight milling of planes and curved surfaces but also gear and thread cutting, drilling, boring and slotting when suitably equipped①. It derives its name from the fact that its two main components are a column that supports the spindle, and a knee (roughly resembling a human's knee) that supports the work table. It is available as either a horizontal or a vertical machine, as illustrated in Figure 5-2-1. In the horizontal version, an *arbor* usually supports the cutter. The arbor is basically a shaft that holds the milling cutter and is driven by the spindle. An overarm is provided on horizontal machines to support the arbor. On vertical knee-and-column machines, milling cutters can be mounted directly in the spindle without an arbor.

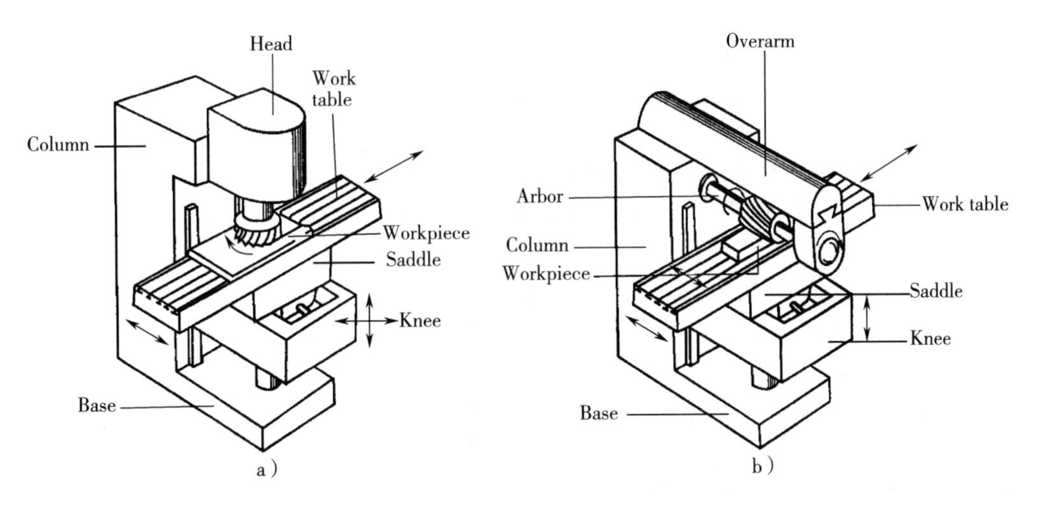

Figure 5-2-1 Two basic types of knee-and-column milling machine
a) Vertical milling machine b) Horizontal milling machine

Bed type (manufacturing) milling machines. In production manufacturing operations, *ruggedness* and the capability of making heavy cuts are of more importance than *versatility*. Bed type milling machines are made for these conditions. The table is mounted directly on the bed and has only *longitudinal* motion. The cutter is mounted in a spindle head that can be adjusted vertically along the machine column. Single spindle bed machines are called simplex mills, and are available in either horizontal or vertical models. Duplex mills use two spindle heads. The heads are usually positioned horizontally on opposite sides of the bed to perform simultaneous operations during one feeding pass of the work②. Triplex mills add a third spindle mounted vertically over the bed to further increase machining capability.

Planer type milling machines. These machines are used only for very large workpiece involving table travels in meters. Through the use of different types of milling heads and cutters, a wide variety of surfaces can be machined with a single setup of the workpiece. This is a great advantage where heavy workpieces are involved.

Special type milling machines. Many types of special milling machines are made to accomplish specific kinds of work more easily than the standard types. In this category are the duplicating or profiling machines, rotary table mills, *drum* type, copy milling (die sinking machines), key way milling machines, and others especially adapted for certain jobs. These machines provide special facilities to suit specific applications that are not *catered* to by the other classes of milling machines.

Questions

1. What are the four types of milling machines?
2. What is a knee-and-column milling machine?
3. Describe the principal characteristics of the basic machine tool for milling.
4. Describe a typical production-type milling machine.
5. What are the relative advantages and disadvantages of knee-and-column and bed type milling machines?

New Words and Expressions

1. orientation [ˌɔːrienˈteiʃən] n. 定向，取向，定位
2. knee [niː] n. 升降台 vt. 用膝盖碰 adj. 直角的
3. column [ˈkɔləm] n. 柱状物，(表格) 栏，立柱
4. arbor [ɑːbə] n. (机床的) 刀轴，心轴
5. rugged [ˈrʌigid] adj. 严格的，粗状的，不平的
6. versatility [ˌvəːsəˈtiləti] n. 多功能性
7. longitudinal [ˌlɔndʒiˈtjuːdinl] adj. 纵向的 n. 纵梁
8. drum [drʌm] n. 鼓，圆筒，线轴
9. cater [ˈkeitə] vi. 投合，迎合
10. the (by) use of 通过利用

11. for the use of 供……应用，应……要求
12. either ... or ... 或……或，不是……就是
13. derive ... from ... 从……衍生，引申
14. cater to 投合，迎合

Notes

[1] This type of machine is capable not only of doing straight milling of planes and curved surfaces but also gear and thread cutting, drilling, boring and slotting when suitably equipped.

在装备完善的情况下，这种类型的机床不仅能够进行平面和曲面的铣削，而且还可以加工齿轮和螺纹、钻孔、镗孔和开槽（插削）。

句中 not only ... but also ... 可译为“不仅……而且……”。

[2] The heads are usually positioned horizontally on opposite sides of the bed to perform simultaneous operations during one feeding pass of the work.

两个主轴头通常水平安装在床身两端，以保证在工件的一次进给行程中可执行同步性操作。

句子中 on opposite sides 意为“在对面，在两侧”。

Glossary of Terms

1. knee-and-column （铣床的）升降台
2. knee-and-column milling machine 升降台式铣床
3. universal knee-and-column milling machine 万能升降台式铣床
4. turret-type knee-and-column milling machine 塔式升降台式铣床
5. hand-feed milling machine 手动进给铣床
6. bed type milling machine 床身式铣床，高刚性铣床
7. rotary-table milling machine 回转式铣床
8. planer type milling machine 定梁龙门铣床
9. special type milling machine 专用铣床
10. copy（profiling，duplicating）milling machine 仿形铣床，靠模铣床
11. key way milling machine 键槽铣床
12. fixed-bed-type milling machine 固定床身式铣床
13. simplex mills（simplex milling machine） 单柱铣床
14. gang milling 多刀铣削
15. up（conventional）milling 逆铣
16. down（climb）milling 顺铣
17. peripheral（plane）milling 圆周（平面）铣
18. surface milling machine 平面铣床
19. face milling 铣面
20. end milling 立铣（端铣）

21. slot milling 铣槽
22. surface machining 曲面加工
23. ball nose mill 球头铣刀
24. vertical milling machine 立式铣床
25. horizontal milling machine 卧式铣床
26. milling cutter 铣刀
27. accessories of milling machine 铣床辅具
28. milling and boring machine 铣镗两用机床
29. single angle milling cutter 单角铣刀
30. arbor-type cutter 套式铣刀

Reading Materials

Milling Operations

Milling is a basic machining process by which a surface is generated progressively by the removal of chips from a workpiece as it is fed to a rotating cutter in a direction perpendicular to the axis of the cutter. In some cases the workpiece remains stationary, and the cutter is fed to the work. In nearly all cases, a multiple-tooth cutter is used so that the material removal rate is high. Common milling operations are depicted in Figure 5-2-2. Often the desired surface is obtained in a single pass of the cutter or work and, because very good surface finish can be obtained, milling is particularly well suited, and widely used, for mass-production work. Unquestionably, more flat surfaces are produced by milling than by any other machining process.

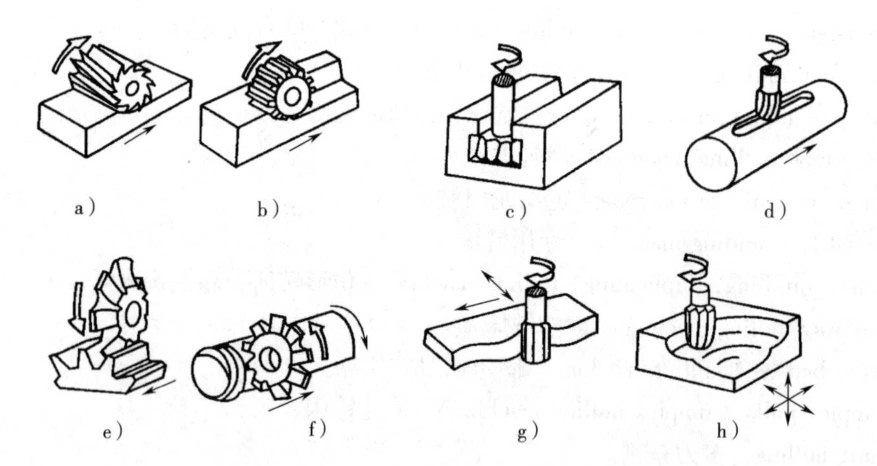

Figure 5-2-2 Common milling operations

a) Slab or plain milling b) Side milling c) Slotting d) End milling
e) Form milling f) Thread milling g) Profile milling h) Surface contouring

Types of milling operations. Milling operations can be classified into two broad categories, peripheral milling and face milling, each having many variations.

In peripheral milling a surface is generated by teeth located on the periphery of the cutter body. The surface is parallel with the axis of rotation of the cutter. Both flat and formed surfaces can be produced by this method, the cross section of the resulting surface corresponding to the axial contour of the cutter. This process often is called slab milling.

In face milling the generated surface is at right angles to the cutter axis and is the combined result of the actions of the portions of the teeth located on both the periphery and the face of the cutter. Most of the cutting is done by the peripheral portions of the teeth, with the face portions providing some finishing action.

Peripheral milling operations usually are performed on machines having horizontal spindles, whereas face milling is done on both horizontal and vertical spindle machines.

The generation of surfaces in milling. In milling, surfaces can be generated by two distinctly different methods, illustrated in Figure 5-2-3. Up milling is the traditional way to mill. Called conventional milling, the cutter rotates against the direction of feed of the workpiece. In climb or down milling the rotation is in the same direction as the feed. The method of chip formation is completely different in the two cases. In up milling the chip is very thin at the beginning, where the tooth first contacts the workpiece, and increases in thickness, becoming a maximum where the tooth leaves the workpiece. In down milling, maximum chip thickness occurs close to the point at which the tooth contacts the workpiece. Because the relative motion tends to pull the workpiece into the cutter, all possibility of looseness in the table feed screw must be eliminated if down milling is to be used. Another advantage of down milling is that the cutting force tends to hold the work against the machine table, permitting lower clamping forces.

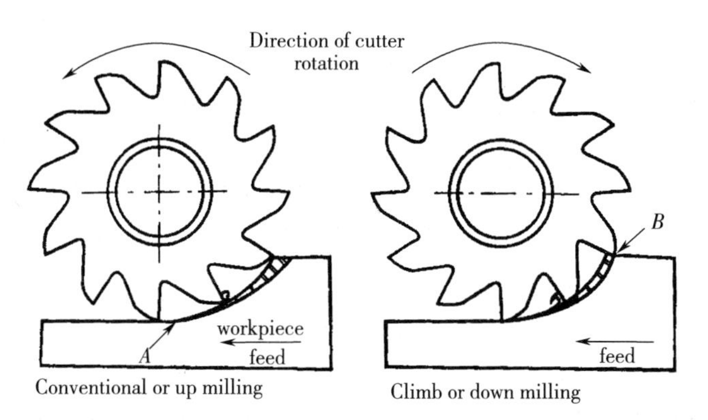

Figure 5-2-3 Difference between conventional and climb milling

Milling Cutters

Flat or curved surface, inside or outside, of almost all shapes and sizes can be machined by milling. The same kind of a surface can often be milled in several ways. For instance, plane surfaces may be machined by slab milling, side milling, or face milling. The method for any specific job may be determined by the kind of milling machine used, the cutter, or the shape of the workpiece and the position of the surface.

Kinds of milling cutters. Many kinds and sizes of cutters are needed for the large variety of work that can be done by milling. Many standard cutters are available, but when they are not adequate, special cutters are made. A cutter is often named for the kind of milling operation it does.

Milling cutters are made of various diameters, lengths, widths and numbers of teeth. Milling cutters may be of the solid, tipped, or inserted tooth types with the same materials as for single-point tools. Large cutters commonly have teeth of expensive material, inserted and locked in place in a soft-steel or cast-iron body, like the face milling cutter.

Plain milling cutters are cylindrical or disk-shaped, have straight or helical teeth on the periphery, and are used for milling flat surfaces. This type of operation is called plain, or slab milling.

Side milling cutters are similar to plain milling cutters except that the teeth extend radially part way across one or both ends of the cylinder toward the center. The teeth may be either straight or helical. Frequently, these cutters are relatively narrow, being disklike in shape. Two or more side milling cutters often are spaced on an arbor to make simultaneous, parallel cuts, in an operation called straddle milling.

Staggered-tooth milling cutters are narrow cylindrical cutters having staggered teeth, and with alternate teeth having opposite helix angles. They are ground to cut only on the periphery, but each tooth also has chip clearance ground on the protruding side. These cutters have a free cutting action that makes them particularly effective in milling deep slots.

Angle milling cutters are made in two types—single-angle and double-angle. Single-angle cutters have teeth on the conical surface, usually at an angle of 45° to 60° to the plane face. The teeth may also extend radially on the larger plain face. Double-angle cutters have V-shaped teeth, with both conical surfaces at an angle to the end faces, but not necessarily at the same angle. The V-angle usually is 45°, 60°, or 90°. Angle cutters are used for milling slots of various angles or for milling the edges of workpieces to a desired angle.

Form milling cutters have the teeth ground to a special shape—usually an irregular contour—to produce a surface having a desired transverse contour. They are cam-relieved and are sharpened by grinding only the tooth face, thereby retaining the original contour as long as the plane of the face remains unchanged with respect to the axis of rotation. Convex, concave, corner-rounding and gear-tooth cutters are common examples.

End mills are shank-type cutters having teeth on the circumferential surface and one end. They thus can be used for facing, profiling, and end milling. The teeth may be either straight or helical, but the latter is more common. Small end mills have straight shanks, whereas taper shanks are used on larger sizes. Plain end mills have multiple teeth that extend only about halfway toward the center on the end. They are used in milling slots, profiling, and facing narrow surfaces. Shell end mills are solid-type, multiple-tooth cutters, similar to plain end mills but without a shank. Hollow end mills are tubular in cross section, with teeth only on the end but having internal clearance. They are used primarily on automatic screw machines for sizing cylindrical stock, producing a short, cylindrical surface of accurate diameter.

Unit 3 Drilling Machines

Text

Drilling machines are made in many forms and sizes. Portable or hand drills are well known. The drilling machines commonly used for precision metal working are known as drill presses. Drilling machines, called drill presses, consist of a base, a column that supports a powerhead, a spindle and a work table. The heart of any drilling machine is its spindle. In order to drill satisfactorily, the spindle must rotate accurately and also resist whatever side forces result from the drilling. The work table on drilling machines may be moved up and down on the column to *accommodate* work of various sizes.

Vertical drill presses. The main *features* of standard upright drill presses are depicted in Figure 5-3-1. A column on a base carries a table for the workpiece and a spindle head. The table is raised or lowered manually, often by an *elevating* screw, and can be clamped to the column for *rigidity*.

A bench-type drilling press is a light machine for small work. Light drill presses, bench and upright, which are hand fed so the operator can feel the resistance met by the drill are called sensitive drill presses. They are advantageous for feeding small drills to avoid *breakage*.

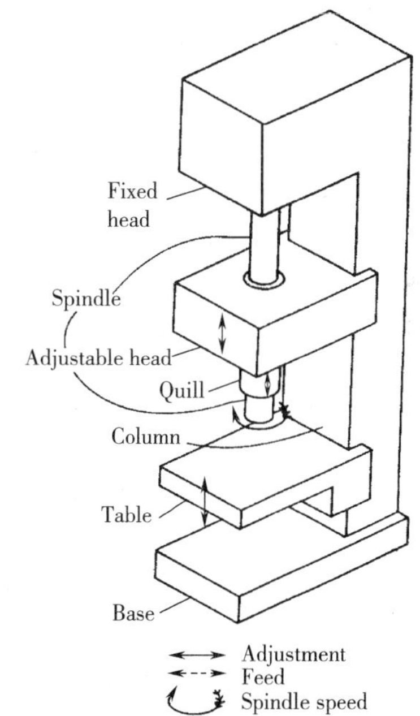

Figure 5-3-1 Principal parts and movements of a single-spindle upright drill press

Gang-drilling machines. In mass production, gang-drilling machines often are used where several related operations, such as holes of different sizes, reaming, or counterboring, must be done on a single part①. These consist essentially of several independent columns, heads and spindles mounted on a common base and having a single table. The work can be slid into position for the operation at each spindle. They are available with or without power feed. One or several operations may be used.

Turret-type drilling machines. Turret-type, upright drilling machines are used where a series of holes of different sizes or a series of operations (such as center drilling, drilling, reaming and spot facing) must be done repeatedly in succession. After the selected tools are set in the turret, each can quickly be brought into position to be driven by the power spindle merely by rotating the turret, rather than requiring moving and positioning of the workpiece, as with a gang-drilling machine[②].

Radial drilling machines. When holes must be drilled at different locations on large workpieces which cannot readily be moved and clamped on an upright drilling machine, the radial drilling machines are employed. These have a large, heavy, round, vertical column supported on a large base. The column supports a radial arm that can be raised and lowered by power and rotated over the base. The spindle head, with its speed and feed changing mechanism, is mounted on the radial arm. It can be moved horizontally to any desired position on the arm.

Plain radial drilling machines provide only a vertical spindle motion. On semi-universal machines, the spindle head can be swung about a horizontal axis normal to the arm to permit the drilling of holes at an angle in a vertical plane. On universal machines, an additional angular adjustment is provided by rotation of the radial arm about a horizontal axis. This permits holes to be drilled at any desired angle.

Multiple-spindle drilling machines. Where a number of parallel holes must be drilled in a part, multiple-spindle drilling machines are used. These are mass production machines with as many as 50 spindles driven by a single powerhead and fed simultaneously into the work. Multiple-spindle drilling machines are available with a wide range of numbers of spindles in a single head, and two or more heads frequently are combined in a single machine. Often drilling operations are performed simultaneously on two or more sides of a workpiece.

Deep-hole drilling machines. Special machines are used for drilling long (deep) holes, such as are found in *rifle barrels*, connecting rods, and long spindles. High cutting speeds, very light feeds, a positive and *copious* flow of cutting fluid to assure rapid chip removal, and adequate support for the long, *slender* drills are required.

Questions

1. What is the heart of any drilling machines used? Explain the function of each type of drilling machines.
2. What are the different types of drilling machines?
3. How does a gang-drilling machine differ from a multiple-spindle drilling machine?
4. What means are used to drill deep holes?

New Words and Expressions

1. accommodate [əˈkɔmədeit] vt. 调节
2. feature [ˈfiːtʃə] n. 特征
3. elevate [ˈeliveit] vt. 举升
4. rigidity [riˈdʒiditi] n. 刚性，刚度
5. breakage [ˈbreikidʒ] n. 破损，破坏，损耗
6. rifle [ˈraifl] n. 步枪，来福枪
7. barrel [ˈbærəl] n. 枪管

8. copious [ˈkəupjəs] adj. 大量的，丰富的
9. slender [ˈslendə] adj. 细长的
10. a series of 一系列，许多
11. bring into use 使用，启用

Notes

[1] In mass production, gang-drilling machines often are used where several related operations, such as holes of different sizes, reaming, or counterboring, must be done on a single part.

在大批量生产时，排式钻床通常用于必须在一个零件上进行多项相关操作的情况，如钻不同尺寸的孔、铰孔或沉孔。

句中由 such as 引出同位语，进一步说明 operations。

[2] After the selected tools are set in the turret, each can quickly be brought into position to be driven by the power spindle merely by rotating the turret, rather than requiring moving and positioning of the workpiece, as with a gang-drilling machine.

在转塔式刀架中的所选刀具确定后，仅仅通过动力轴驱动转塔刀架的旋转，且每一次都能使刀具快速进入位置，而不需要工件的运动和定位，这就如同排式钻床的情况一样。

句中 rather than 可译为“而不”；as with 可译为“如同……的情况一样”。

Glossary of Terms

1. drilling machine 钻床
2. gang drill 排式钻床，排式钻头
3. radial drilling machine 摇（旋）臂钻（床）
4. twist drill 麻花钻
5. gun drilling 深孔钻
6. lathe drill 卧式钻床
7. vertical drill press (upright drilling machine) 立式钻床
8. drill chuck 钻夹头，钻卡
9. drill grinder 钻头磨床
10. drill holder 钻套
11. drill jig 钻模，钻夹具
12. drill sleeve 钻头套筒
13. drill stock 钻柄
14. drill 钻头
15. straight-shank drill 直柄钻头
16. horizontal deep-hole drilling machine 卧式深孔钻床
17. accessories of drilling machine 钻床辅具

18. fixture of drilling machine 钻床夹具
19. multiple-spindle drilling machine 多轴钻床
20. computer numerical control drilling machine 计算机数控钻床
21. turret-type drill press 塔式钻床
22. drilling machines bench 钻床工作台
23. gun drilling machine 深孔钻床，枪孔钻床
24. bench-type drilling press 台式钻床
25. through hole (blind hole) 通孔（不通孔）
26. chamfer, fillet angle 倒角，圆角
27. pin hole 销孔

Reading Materials

Drilling Operations

The kinds of operations to be considered at this time are those concerned mostly with the opening, enlarging, and finish cutting of holes from a small fraction of a millimeter to hundreds of millimeters in diameter. The tools, and not the workpieces, are revolved and fed into the material in most of the operations.

In manufacturing, it is probable that more internal cylindrical surfaces—holes—are produced than any other shape, and a large proportion of these are made by drilling. Consequently, drilling is a very important process. Drilling machines are used for drilling holes, reaming, and other general purposes illustrated in Figure 5-3-2.

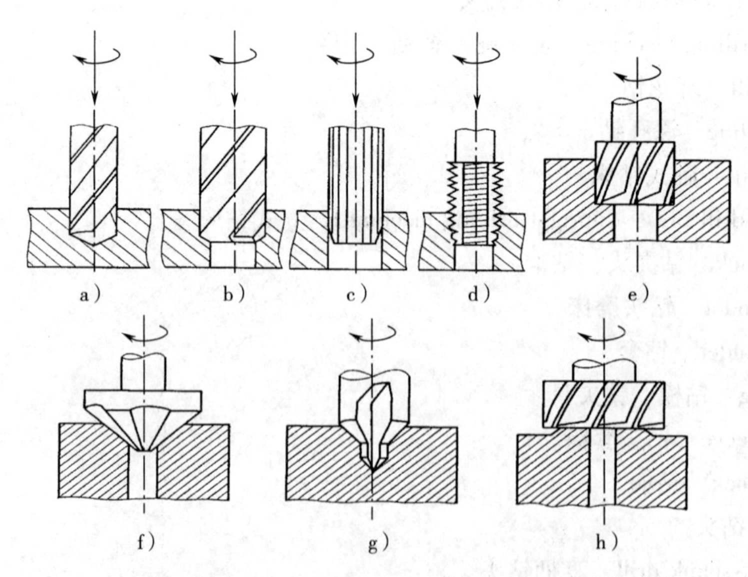

Figure 5-3-2 Machining operations related to drilling

a) Drilling b) Core drilling c) Reaming d) Tapping
e) Counterboring f) Countersinking g) Center drilling h) Spotfacing

Drilling is the easiest way to cut a hole on solid metal. It is done to enlarge holes and then may be called core drilling or counter drilling. Reaming is used to slightly enlarge a hole, to provide a better tolerance on its diameter, and improve its surface finish. The tool is called a reamer, and it usually has straight flutes. Tapping is used to provide internal screw threads on an existing hole. Counterboring provides a stepped hole, in which a large diameter follows a smaller diameter partially into the hole. A counterbored hole is used to seat bolt heads into a hole so the heads do not protrude above the surface. Countersinking is similar to counterboring, except that the step in the hole is cone-shaped for flat head screws and bolts.

Centering is also called center drilling and is used to help start a hole and guide the drill for regular drilling. Spotfacing is similar to milling, which is used to provide a flat machined surface on the workpart in a localized area.

Drills

Types of drills. The most common types of drills are twist drills, which have three basic parts: the body, the point and the shank, shown in Figure 5-3-3. The body contains the two or more spiral or helical grooves, called flutes, separated by lands. To reduce the friction between the drill and the hole, each land is reduced in diameter except at the leading edge, leaving a narrow margin of full diameter to aid in supporting and guiding the drill and thus aiding in obtaining an accurate hole. The lands terminate in the point, with the leading edge of each land forming a cutting edge. The flutes serve as channels through which the chips are withdrawn from the hole and coolant gets to the cutting edges. Although most drills have two flutes, some have three, and some have only one.

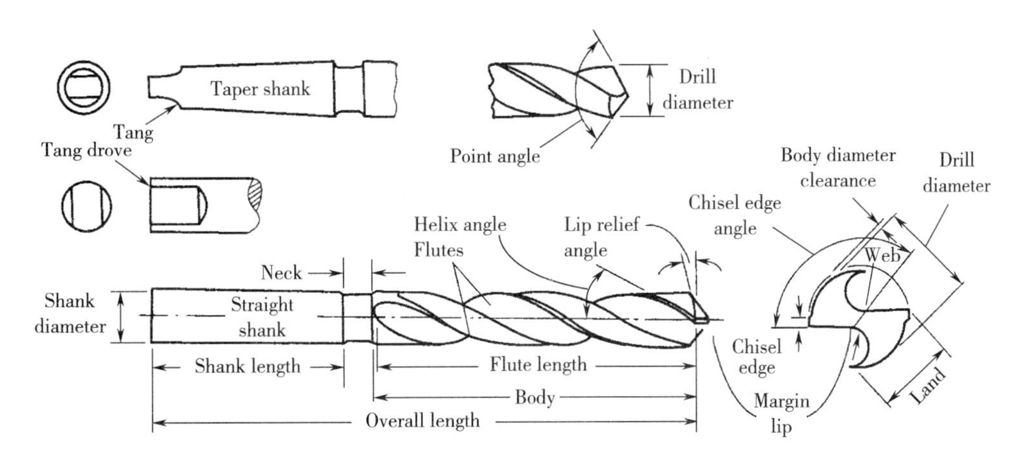

Figure 5-3-3 Standard twist drill elements

Drill shanks are made in several types. The two most common types are the straight and the taper. Straight-shank drills are usually used for sizes up to 13mm and must be held in some type of drill chuck. Taper shanks are available on drills from 3mm to 13mm and are common on drills above 13mm.

Deep-hole drills. Deep holes (over about 10 diameters in depth) are difficult to drill. Chips are difficult to get out of the hole. It is not easy to get the fluid to the point of the drill, and the drill tends to run out too much. These conditions must be corrected in drilling such parts as gun barrels, crankshafts, camshafts and hollow spindles. Certain procedures help, such as revolving both workpiece and drill and withdrawing the drill often, but specific drills for the purpose are often necessary. Deep-hole drills, which contain passages through which coolant is forced to the cutting edges, and which also aid in pushing chips back out of the hole, are used when deep holes are to be drilled.

In recent years, new drill point geometries have resulted in improved hole accuracy, longer life, self-centering action and increased feed-rate capabilities. However, virtually 99% of the drills manufactured have the conventional point. It is often left to the user to regrind the drills to fit his application.

Unit 4 Shapers and Planers

Text

Shapers

Horizontal shapers. A horizontal shaper depicted in Figure 5-4-1 has a horizontal *ram* that reciprocates at cutting speed. A cutting tool is carried on the *toolhead* on the front of the ram. The length and position of the *stroke* of the ram are adjustable so that the tool can be set to cover any part of the maximum stroke of the shaper and need not travel any more than necessary for each job①. The slide that carries the tool on the toolhead can be *swiveled* and clamped at any angle in a vertical plane on the front of the ram.

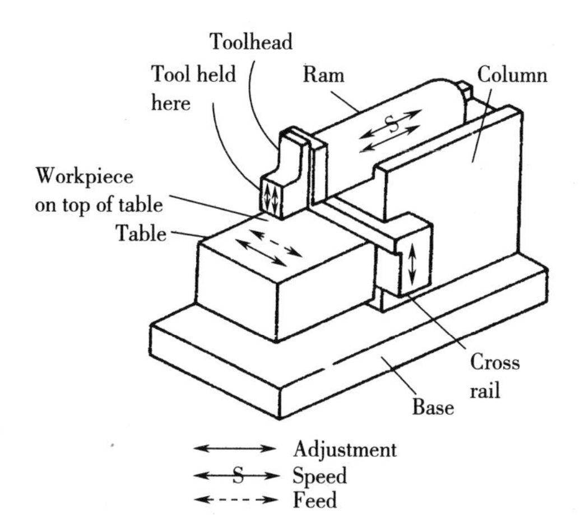

Figure 5-4-1 Principal parts and movements of a horizontal shaper

A cross rail mounted on ways on the front of the column of a horizontal shaper is adjustable up or down. The work table rides *crosswise* on the rail and can be adjusted or fed manually or automatically. Feed takes place just before the beginning of each stroke. The power feed is activated by the ram driving mechanism.

Vertical shapers. A vertical shaper or slotter has a vertical ram and normally a rotary table as depicted in Figure 5-4-2. On some machines the ram is inclinable up to 10° from the vertical, which is useful for cutting inclined surfaces. The rotary table can be *indexed* accurately or rotated continuously by hand or power. It moves on a *saddle* that slides on the bed and has horizontal adjustments and feeds in two directions.

The horizontal table of the vertical shaper is easy to load. The circular plus the two straight line motions of the table permit easy machining of circular, *convex*, *concave* and other curved surfaces②. Slots or *grooves* can be accurately spaced around a workpiece.

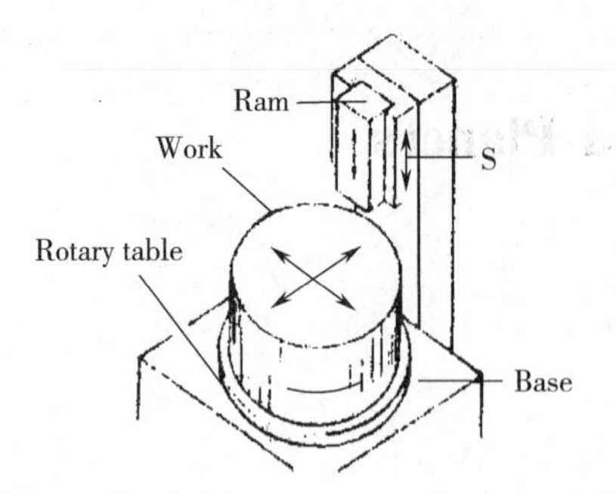

Figure 5-4-2 Schematic of vertical shaper

Planers

The planer or planing machine carries the work on a *massive* table, fully supported by a heavy bed, like the one in Figure 5-4-3, and capable of sustaining heavy loads. The table slides on ways on the bed. Most planners cut in one direction; some in both directions. Tee slots and holes are provided for *bolts*, keys and pins for holding and locating workpieces on the finished table top.

The common type of planer illustrated in Figure 5-4-3 has two heavy housings, also called columns and uprights, at about the middle of the bed, with one on each side of the table. These carry the horizontal cross-rail that can be raised or lowered and then clamped in place. The columns and rail carry the toolheads. Those on the side slide up and down; those on the cross rail slide horizontally.

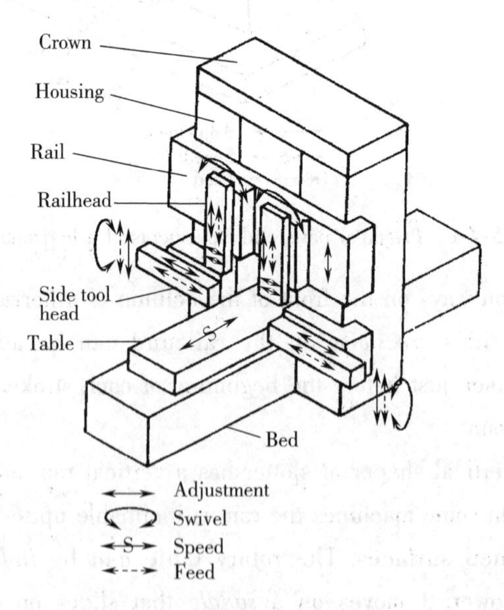

Figure 5-4-3 A double-housing planer

Some planers have hydraulic drives, with locked variable-delivery circuits. Others have mechanical drives, commonly with variable-speed, reversing DC motors. The power is carried from the motor through a set of reduction gears. The last gear in the series *meshes* with a rack fastened to the underside of the table. Table speeds range from 15m/min or less to as much as 90m/min (50fpm to 300fpm).

Questions

1. How are shaping and planing alike, and how do they differ?
2. What are the advantages and disadvantages of shaping and planing compared to other operations?
3. Describe a horizontal shaper and a vertical shaper.
4. For what type of work are shapers best suited?
5. What is the basic difference between a shaper and a planer?

New Words and Expressions

1. ram [ræm] n. 滑枕
2. toolhead [ˈtuːlhed] n. 刀架，刀夹
3. stroke [strəuk] n. 行程
4. swivel [ˈswivl] vt. 旋转
5. crosswise [ˈkrɔswaiz] adv. 交叉地，成十字形地
6. indexed [ˈindekst] adj. 分度的
7. saddle [ˈsædl] n. 滑板，座板
8. convex [ˈkɔnveks] n. 凸面，凸圆体 adj. 凸的，凸起的
9. concave [kɔnˈkeiv] n. 凹面 adj. 凹的，凹入的
10. groove [gruːv] n. 槽，凹槽
11. massive [ˈmæsiv] adj. 粗大的，重的
12. bolt [bəult] n. 螺栓，螺钉
13. mesh [meʃ] vt. 啮合
14. curved surface 曲面
15. concavo-concave 双凹的，两面凹的
16. concavo-convex 一面凹一面凸的，凹凸的
17. variable-delivery circuit 变流量回路
18. on the front of 在……前面，在……前方
19. a set of 成套，批量

Notes

[1] The length and position of the stroke of the ram are adjustable so that the tool can be set to cover any part of the maximum stroke of the shaper and need not travel any more than necessary for each job.

滑枕行程的位置和长度是可调节的，以便安装的刀具可作用于刨床最大行程范围内可能

涉及的任意零件，并且对于要完成的每一项加工不需要任何非必要的移动。

句中由 so that 引导目的状语从句，可译为“为了……，以便……”。

[2] The circular plus the two straight line motions of the table permit easy machining of circular, convex, concave and other curved surfaces.

工作台的圆周运动加两个直线运动，使圆形表面、凸圆体、凹形面以及其他曲面易于加工。

句中 the circular plus the two straight line motions of the table 为主语，可译为“工作台的圆周运动加两个直线运动”。

Glossary of Terms

1. double-housing planer 双柱式龙门刨床
2. open-side planer 单柱刨床，单臂龙门刨床
3. horizontal shaper 卧式刨床
4. vertical shaper 立式刨床
5. plate edge planer 刨边机，板材刨边机
6. horizontal boring machine 卧式镗床
7. horizontal planer (planing machine) 卧式刨床
8. fixture of planning machine 刨床夹具
9. accessories of planing machine 刨床辅具
10. table planing machine 龙门刨床
11. shaping machine 牛头刨床，成形机
12. cross rail 横梁，横导轨
13. step by step variable gear 多级变速齿轮
14. plane planer 平面刨床
15. planer fixture 刨削夹具
16. broaching machine 拉床
17. sawing machine 锯床
18. flat (round, square, triangular) file 扁（圆，方，三角）锉
19. slotting machine (slotter) 插床，立刨床
20. terminology for metal-cutting machine tools 金属切削机床术语
21. modular machine tools 组合机床
22. boring machine 镗床
23. NC jig boring machine 数控坐标镗床
24. machine tool accessory 机床辅件
25. slabbing cutter 刨刀，阔面铣刀
26. gear shaper 刨齿机
27. crank shaper 曲柄牛头刨床

Reading Materials

Shaping and Planing Operations

Both shaping and planing are intended primarily for flat surfaces, horizontal, vertical, or at an angle as indicated in Figure 5-4-4, but can be arranged for machining curved surfaces and slots. Short internal surfaces like square or splined holes can also be shaped.

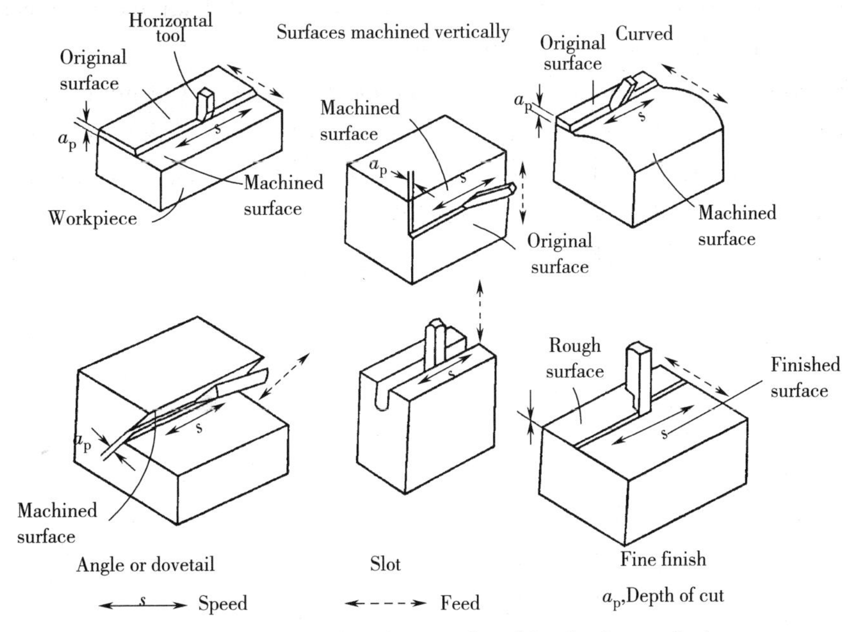

Figure 5-4-4 Typical surfaces machined by shaping or planing

Shaping and planing differ with respect to action and workpiece size. In shaping, the tool is reciprocated over the workpiece surface, while in planing, the workpiece is moved past the tool, in both cases at cutting speeds. In shaping a horizontal surface the workpiece is fed, but for shaping vertical and inclined surfaces and in all planing, the tool is fed an increment for each stroke in a direction perpendicular to the cutting speed. Because of the necessary structure of the machine, shaping is limited to small and moderate-size workpieces.

In most shaping and planing operations, cutting is done in one direction only. The return stroke represents lost time. Thus these processes are slower than milling and broaching, which cut continuously. On the other hand, shaping and planing use single-point tools that are less expensive, are easier to sharpen, and are conductive to quicker setups than the multiple-point tools of milling and broaching. This makes shaping or planing often economical to machine one or a few pieces of a kind.

Types of Planers

The standard or double-housing planer that has been described is capable of heavy service. An open-side planer has a housing only on one side of the bed. The other side is open so that extra wide work can hang over the side of the bed sometimes supported on an auxiliary rolling table alongside. Small sizes of open-side planers are also called shaper planers. A divided or latching table planer has a table in two sections. They may be coupled together for long work or separated so the work on one table may be set up while that on the other is machined.

The planer-type milling machine is a cross between the planer and milling machine. A planer-type grinder has a grinding head in place of one or more of the conventional toolheads on the rail.

A plate planer is a special purpose machine for squaring and beveling the edges of steel plates for armor, ships, or pressure vessels. One edge of the plate is held down and machined by a tool traversed on a carriage along the front of a machine.

When workpieces are too heavy or bulky, the tools may be moved more readily than the work. That is the basis for the design of the pit planer on which the workpiece is mounted on a stationary table. Columns carrying the cross rail and toolheads ride on long ways on both sides of the table.

Unit 5 Grinding Machines

Text

Grinding machines utilize grinding wheels and may be classified according to the types of operations. Thus the broad classes are precision and non-precision grinders. The main types of precision grinding machines are external and internal cylindrical grinding machines and surface grinding machines. Certain types have been developed to do specific operations and are classified accordingly.

Space is available to describe only the basic grinding machines. Most types are available in simple hand-operated models and in various semi- and fully automatic models, as are other machine tools.

Cylindrical center-type grinders. Cylindrical center-type grinders are used for grinding straight and tapered round pieces, round parts with curved lengthwise profiles, fillets, shoulders and faces. A simple and the original form of cylindrical grinder is a head that revolves a grinding wheel and is mounted on the cross-slide or compound rest of a lathe①.

A workpiece is usually held between dead centers and rotated by a dog and driver on the faceplate on a plain cylindrical center-type grinder. The centers are held in the *headstock* and *footstock* of the machine, and do not revolve because that provide the most rigid work support and accuracy between centers. The headstock and footstock are carried on an upper or swivel table and can be positioned to suit the length of the workpiece. The upper table can be swiveled and clamped on a lower table to adjust for grinding straight or tapered work as desired. The amount of taper that can be ground in this way is normally less than 10° because the table bumps the wheel-head if swiveled too far. The lower table slides on ways on the bed of the machine to *traverse* the work past the grinding wheel and can be moved by hand or power.

***Centerless* grinders.** An external cylindrical centerless grinding machine revolves a workpiece on top of a workrest blade between two abrasive wheels. The grinding wheel removes material from the workpiece. The workpiece has a greater *affinity* for and is driven at the same surface speed as the regulating wheel, which is normally a rubber bonded abrasive wheel that turns at a surface speed of 15 m/min to 60 m/min②.

Internal grinders. A chucking internal grinder holds the workpiece on a faceplate or in a chuck or fixture and rotates it around the axis of the hole ground. The revolving grinding wheel is reciprocated lengthwise through the hole and is fed crosswise on a slide to engage the workpiece. The workhead, and sometimes the *wheelhead*, can be swiveled to adjust for straight and tapered holes and faces.

Surface grinders. Surface grinding is concerned primarily with grinding plane or flat surfaces but also is capable of grinding irregular, curved, tapered, convex and concave surfaces.

Conventional surface grinders may be divided into two classes. One class has reciprocating tables for work ground along straight lines. This type is particularly suited to pieces that are long or have stepped or curved profiles at right angles to the direction of grinding. The second class covers the machines with rotating work tables for continuous rapid grinding.

Surface grinders may also be classified according to whether they have horizontal or vertical grinding wheel spindles. Grinding is normally done on the *periphery* of the wheel with a horizontal spindle. The area of contact is small, and the speed is uniform over the grinding surface. Small-grain wheels can be used, and the finest finishes obtained. Grinding with a vertical spindle is done on the side of the wheel, which may be solid, sectored, or *segmental*. The area of contact may be large, and stock can be removed rapidly.

Questions

1. What are the main types of precision grinding machines?
2. Describe a plain cylindrical center-type grinder and tell what it does.
3. Describe the action of a centerless grinder. What are its advantages?
4. What are the four classes of surface grinders?
5. What is the advantage of each kind of surface grinder?

New Words and Expressions

1. headstock [ˈhedstɔk] n. 头架
2. footstock [ˈfutstɔk] n. 尾架，尾座，顶尖座
3. traverse [ˈtrævə（ː）s] vt. 横过，穿过，在……来回移动
4. centerless [ˈsentəlis] adj. 无中心的
5. affinity [əˈfiniti] n. 密切关系，类似，近似
6. wheelhead [ˈwiːlhed] n. 砂轮头，磨头
7. periphery [pəˈrifəri] n. 圆周，周围
8. segmental [segˈmentəl] adj. 部分的
9. crisscross [ˈkriskrɔs] adj. 十字形的，交叉的 n. 十字形
10. scratch [skrætʃ] n. 刻痕，划痕
11. on the side of 站在……一边，帮着……
12. have an affinity for 被吸引，喜爱

Notes

[1] A simple and the original form of cylindrical grinder is a head that revolves a grinding wheel and is mounted on the cross-slide or compound rest of a lathe.

外圆磨床的简单和最初的形式是一个旋转的砂轮头安装在车床的横向刀架或组合刀架上的。

句中 is mounted on 可译为“把……安装在……上”；that revolves a grinding wheel 为定语从句，修饰 a head。

[2] The workpiece has a greater affinity for and is driven at the same surface speed as the regulating wheel, which is normally a rubber bonded abrasive wheel that turns at a surface speed of 15 m/min to 60 m/min.

工件和导轮之间的摩擦力带动工件以与导轮相同的表面速度转动，导轮通常是含有橡胶粘接剂的砂轮，并以 15 ~60 米/分钟的表面速度旋转。

句子中 has a greater affinity for 意为“与……结合紧密”，意指导轮依靠摩擦力带动工件转动；而 which is normally a rubber bonded abrasive wheel that turns at a surface speed of 15 m/min to 60 m/min 则作为定语从句修饰 the regulating wheel。

Glossary of Terms

1. precision grinding machine 精密磨床
2. precision internal grinder 精密内圆磨床
3. external cylindrical grinding machine 外圆磨床
4. internal cylindrical grinding machine 内圆磨床
5. face grinding machine (grinder) 平面磨床
6. slider grinder 导轨磨床
7. external cylindrical centerless grinding machine 外圆无心磨床
8. grinding wheel 砂轮，磨轮
9. wheel depth of cut 砂轮切削深度
10. grinding wheel wear 砂轮磨损
11. face (plane, surface) grinding 平面磨削
12. face grinding wheel 平面磨轮
13. internal grinding 内圆磨削
14. external grinding 外圆磨削
15. abrasive cloth 砂布
16. chucking internal grinder 卡盘式内圆磨床
17. conventional surface grinder 普通平面磨床
18. rotary-table surface grinder 回转式平面磨床
19. fixture of grinding machine 磨床夹具
20. accessories of grinding machine 磨床辅具
21. universal tool grinding machine 万能工具磨床
22. form grinding machine 成形磨床
23. copy (contour) grinding machine 仿形磨床
24. honing machine 珩磨机
25. spline shaft grinding machine 花键轴磨床

26. electromagnetic chuck 电磁吸盘
27. centreless grinding 无芯磨削

Reading Materials

Grinding Operations

The wide variety of grinding operations can be classified by the main types of surface to be ground. These include cylindrical, flat and internal surfaces.

Cylindrical grinding. As can be seen in the schematic drawing of Figure 5-5-1, there are two distinct types of external cylindrical grinders; those that hold the workpiece between centers or in a chuck, and those in which the workpiece is rotated between two wheels, or centerless grindings.

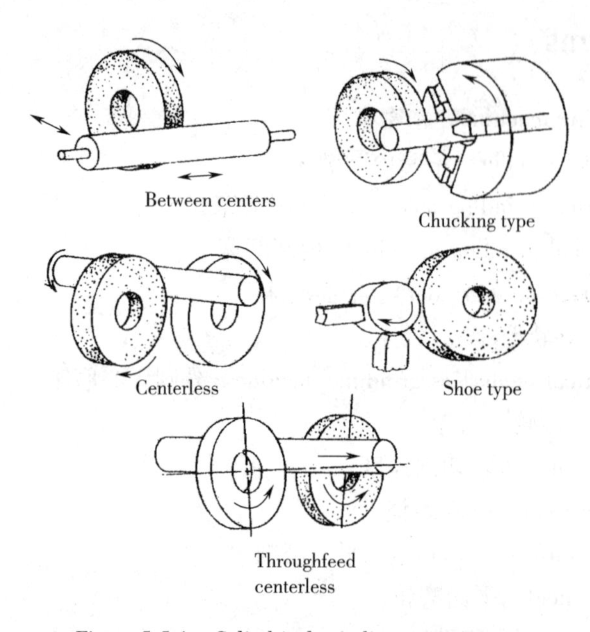

Figure 5-5-1 Cylindrical grinding operations

As the name implies, cylindrical grinding is used to grind cylindrical parts that may be straight, stepped, or tapered. The workpiece is driven by an independent motor and rotates opposite to the wheel rotation at the point of contact.

Centerless grinding. In centerless grinding, the workpieces are held between the grinding wheel, a regulating wheel, and workrest blade. By tilting the rotational axis of the regulating wheel with respect to the grinding wheel, the workpiece is given a longitudinal movement or throughfeed. Once in contact with the grinding wheel the workpiece automatically feeds to the far side of the grinding wheel. Infeed grinding is much the same as plunge grinding and is used on workpieces that have a shoulder larger than the ground diameter. The work is inserted and either the grinding wheel or the regulating wheel is fed in a preset distance to obtain the proper diameter. The wheels are then separated and the piece is either removed by hand or by an automatic ejector.

Surface grinding. All grinding can be construed to be surface grinding. However, the term here refers to flat surfaces. Many types of machines are available for grinding flat surfaces depending upon the workpiece size and shape. The principal types are shown schematically in Figure 5-5-2. The machines can be broadly divided into horizontal spindle and vertical spindle and further subdivided as reciprocating table and rotary table.

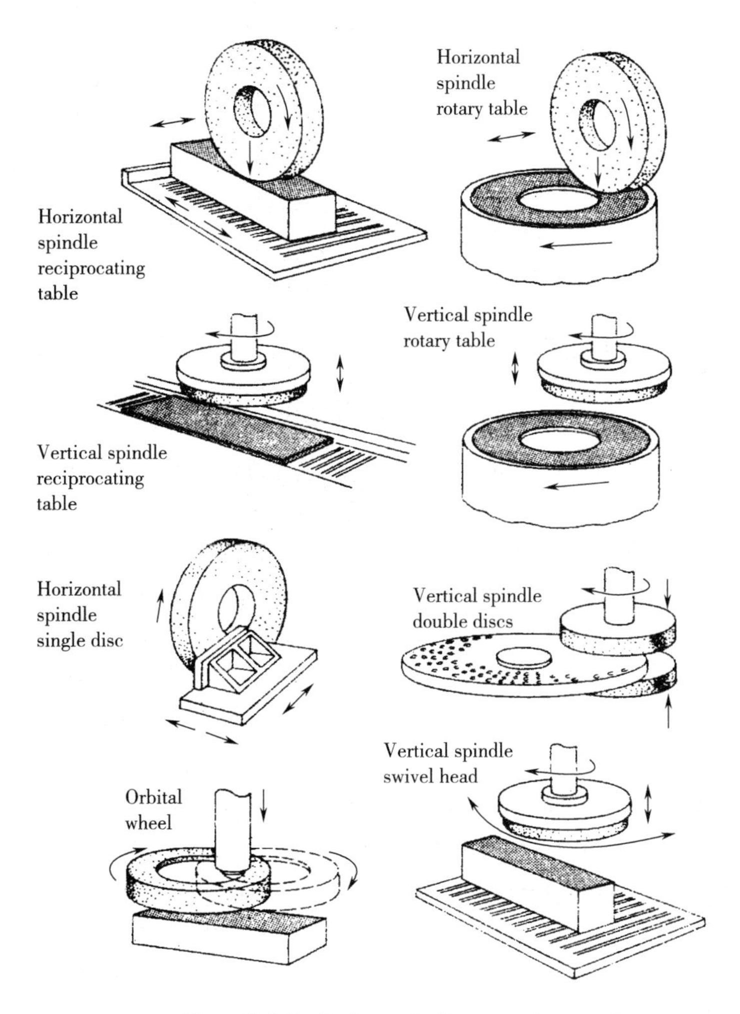

Figure 5-5-2 Surface grinding operations

The most common surface grinder is the horizontal type with a high-speed reciprocating table. In this type the work is typically held on an electromagnetic chuck.

Internal grinding. Internal grinding is used to obtain very accurate size and good finish on holes made by other operations. Machines made that are able to finish holes from a few thousandths of an inch in diameter to 5 feet or more. The basic types of internal grinders are shown in Figure 5-5-3; they are the chucking and centerless types. A variation of the centerless, faceplate/shoe type is also shown.

The chucking type machine is the most common type for internal grinding. Chucks may include four jaw chucks, collet chucks, faceplates and magnetic chucks. One problem often associated with internal grinding is the pressure of the chuck jaws in causing the part to distort.

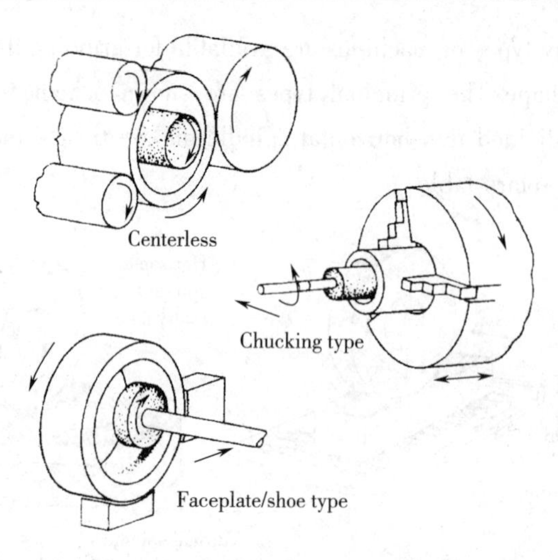

Figure 5-5-3 The basic types of internal grinders

Grinding Wheels

Grinding, wherein the abrasives are bonded together into a wheel of some shape is the most common abrasive machining process. The operation of grinding wheels is greatly affected by the bonding material and the spatial arrangement of the abrasive particles, known as the structure.

Grinding wheel structure. The spacing of the abrasive particles with respect to each other is called structure. Close-packed grains have dense structure, while open structure means widely spaced grains. Open-structure wheels have greater chip clearance but obviously fewer cutting edges per unit area. A typical vitrified grinding wheel will consist of 50% (volume percentages) abrasive particles, 10% bond, and 40% voids or pores.

Bonding materials for abrasive wheels. Bonding material is a very important factor to be considered in selecting a grinding wheel. It determines the strength of the wheel, thus establishing the maximum speed at which it can safely be operated. It determines the elastic behavior or deflection of the grits in the wheel during grinding. The wheel can be hard or rigid or it can be flexible. Finally, the bond determines the force that is required to dislodge an abrasive particle from the wheel and thus plays a major role in the cutting action. Five types of bonding materials are in common use.

Grinding wheel selection. If optimum results are to be obtained, it is very important that the proper grinding wheel be selected for a given job. Several factors must be considered. Probably the first is the shape size of the wheel. Obviously, the shape must permit proper contact between the wheel and all of the surfaces that must be ground. Grinding wheel shapes have been standardized by the Grinding Wheel Manufacturers' Association. The size of the wheel to be used is determined, primarily, by the spindle speeds available on the grinding machine and the proper cutting speed for the wheel, as dictated by the type of bond. Other factors which will influence the choice of wheel to be selected will include the workpiece material, the amount of stock to be removed, the shape of the workpiece, and the accuracy and surface finish desired.

Chapter 6 Machining Processes

Unit 1 Manufacturing Processes

Text

Process of production. Manufacturing is the utilization and management of materials, people, equipment and money to produce products. For successful and economical manufacturing to be achieved, planning must start at the design stage and continue through the selection of materials, processes, equipment, and the *scheduling* of production. These related processes constitute the so-called production process. It is usually completed by the co-ordinated work of different factories or workshops. To do so can take advantage of specialized manufacturing processes and make production simple[1]. It also has the effects of raising production efficiency, *guaranteeing* the quality of products and reducing the cost. Many products, such as sewing machine and automobiles, involve specialized manufacturing.

Process of part production. The process by which the shape, dimensions and performance of raw material or rough casting are changed directly by mechanical manufacturing into desired parts is called component manufactured part production, or process *route*[2].

When parts are assembled into a product, it is called the assembly process.

Part production. Part production consists of one or several operations in sequence. A rough casting can be turned into a finished or semi-finished product by the operations.

Operations. The term operation is defined as when one operator (or a group of operators) works in a definite position (one machine tool or *clipping* table) to achieve the continuous manufacturing of one or several parts simultaneously. An operation is the basic element of process procedure as well as of the production schedule and of cost *estimation*.

Figure 6-1-1 is the engineering drawing of a shaft. If the number of the parts in a *batch* is small,

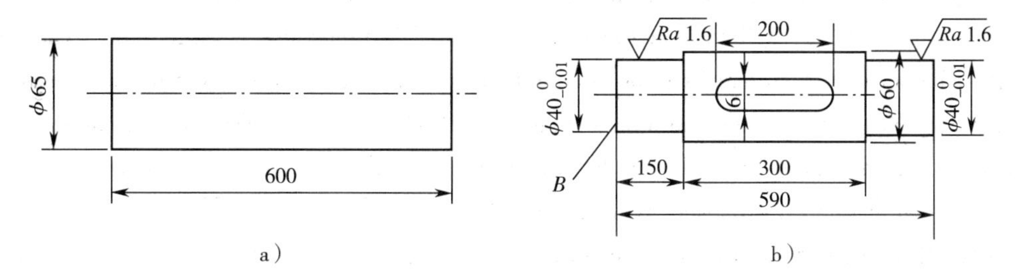

Figure 6-1-1 The drawing of a shaft

a) The stock b) Finished product

the manufacturing process can be completed through five operations, which are shown in Table 6-1-1. The stock shown in Figure 6-1-1a will be turned into the shaft shown in Figure 6-1-1b through the five operations *sequentially*.

Table 6-1-1 The process of shaft manufacturing (Small batch)

Operation No.	Operation name	Machine tool
1	Face ending, center drill ending	Engine lathe
2	Cylindrical turning	Engine lathe
3	Key way milling	Upright milling machine
4	Cylindrical grinding	Grinding machine
5	Deburr	Clipping table

For the manufacturing of the same part as shown in Figure 6-1-1, if the number of the parts is large, the operation 1 in Table 6-1-1 can be divided into two operations, facing and centering one end in an engine lathe by one operator, and then *transferring* to another engine lathe for facing and centering another end by another operator. As the two ends and center holes are not manufactured at the same time, by the same operator or in the same machine tool, the operation 1 has now become two operations according to the definition of an operation.

Operation step. An operation step is the basic element of any operation. One operation step is defined as an operation that is completed under the condition when the surface to be manufactured, the cutting parameters (including cutting speed and feed) and the cutting tool remain the same. When any of the cutting parameters is changed, it will become another operation step. When several surfaces of a part are manufactured simultaneously, it is called a composite operation step.

The purpose of classifying an operation into steps is to make a complex operation more *convenient* and easy to analyze and describe, and to make it more efficient to organize the manufacturing and calculate the operation time.

Cutting pass. When the amount of metal to be removed is large, several cutting passes may need to be carried out. A cutting pass is a kind of operation step that is completed by a single type of tool to cut the same surface under the same cutting parameters.

Setup. Before the part can be machined, it must have a proper position relative to the cutting tool, and be clamped by a fixture in the machine tool. This procedure is called the fixturing. The operation completed through one fixturing is called single setup. Setup is one part of the process. Each operation may have either single setup or several setups. In the same operation, the number of setups should be arranged to be as few as possible, which not only can the production efficiency be raised, but also the production error caused by setups can be reduced[③].

Operation position. In order to reduce the number of setups, a rotational worktable or fixture is usually used to make sequential machining in different positions possible from a single setup. The job performed in one position is called one operation position.

Questions

1. What is the mechanical technical process, or mechanical process route?
2. What is an operation step?
3. Explain the principle of arranging the machining sequence.
4. What is a cutting pass?
5. What is single setup?

New Words and Expressions

1. schedule [ˈʃedjuːl] n. 时间表，计划表，进度表 vt. 将……列入计划，安排
2. guarantee [ˌɡærənˈtiː] n. 保证 vt. 保证，担保
3. route [ruːt] n. 路线，路程 vt. 给……定路线，安排……的程序
4. clipping [ˈklipiŋ] n. 剪断，限制 adj. 快速的
5. estimation [ˌestiˈmeiʃən] n. 估计，评价
6. batch [ˈbætʃ] n. 一次生产量，一批，批量
7. sequential [siˈkwinʃəl] adj. 连续的，有顺序的
8. transfer [trænsˈfəː] vt. 转移，转动，传递
9. convenient [kənˈviːnjənt] adj. 方便的，便利的
10. be carried out 实行，进行，执行
11. be arranged to 按……排列，按……布置
12. according to schedule 按照预定计划
13. on schedule 准时，按照预定时间

Notes

[1] To do so can take advantage of specialized manufacturing processes and make production simple.

这样做，能够发挥专业化制造工艺的优点，并简化生产过程。

句中 to do so 可译为“这样做”；由 and 连接两个并列句。

[2] The process by which the shape, dimensions and performance of raw material or rough casting are changed directly by mechanical manufacturing into desired parts is called component manufactured part production, or process route.

通过机械加工方法，直接使原材料或毛坯铸件的形状、尺寸和性能变为符合目标零件要求的过程，被称为零部件生产，即机械加工工艺过程。

句中 by which the shape, dimensions and performance of raw material or rough casting are changed directly by mechanical manufacturing into desired parts 为限定性定语从句，修饰 the process（在句中做主语）。

[3] In the same operation, the number of setups should be arranged to be as few as possible, which not only can the production efficiency be raised, but also the production error caused by

setups can be reduced.

在同一工序内，安装的次数应安排得尽可能少，这样不仅能提高生产效率，而且还能减少由于安装引起的加工误差。

句中 the number of 意为“……的数目”; as few as possible 可译为“尽可能少”; 由 which 引出非限定性定语从句，修饰前面整个句子。

Glossary of Terms

1. production process 生产过程
2. specialized manufacturing 专门制造，专业化制造
3. process route 工艺过程
4. assembly process 装配过程
5. operation (process) step 工步
6. operation position 工位
7. operation 工序
8. cutting parameter 切削参数
9. composite operation step 复合工步
10. operation sheet 工艺规范，操作说明书
11. process sheet (chart) 工艺卡片
12. cutting pass 走刀
13. single setup 单次安装
14. several setups 多次安装
15. rotational worktable 旋转工作台
16. sequential machining 顺序加工
17. one operation position 单工位
18. working position 工况，工作状态
19. single production 单件生产
20. batch production 批量生产
21. batch process 成批量生产
22. large quantity production 大批量生产
23. production runs 流水线生产
24. process flow diagram (sheet) 工艺流程图
25. manufacturing operation 加工工序
26. processing parameter 工艺参数

Reading Materials

The Function and Design Method of Process Planning File

The term process route is the process planning file in which the process and operation method are specified. In another words, the aligned sequence of each operation, the part dimensions, tolerance and technical requirement, technical equipment and measures, cutting parameters, standard production time, and class of operators' skill are all included in the process planning file.

The format of process planning sheet. Operation sheets vary greatly as to details. The simpler types often list only the required operations and the machines to be used; speeds and feeds may be left to the discretion of the operator, particularly where skilled workers and small quantities are involved. However, it is common practice for complete details to be given regarding tools, speeds, and often the time allowed for completing each operation. Such data are necessary if the work is to be done on NC machines, and experience has shown that these preplanning steps are advantageous where ordinary machine tools are used.

The function of process planning sheet. ①The main technique for giving regulations to guide the production. ②The basis of organization and management. ③The main information to guide the building, enlarging or rebuilding of a mechanical factory.

According to the process planning sheet, the kinds, type and number of machine tools, the production area needed, the plan arrangement of equipment, the number, type and class of operators required can all be decided. Then, the planning of factory preparation, enlarging or rebuilding will be decided accordingly.

In general, the process planning file is a necessary technical file for every mechanical manufacturing factory or workshop. It can be used to make preparation before manufacturing, to guide the operation during production and to inspect the production. Therefore, each operator, techical engineer or manager in the factory or workshop must follow the process planning sheet to organize the production in order to assure the product quality, raise production efficiency, reduce the cost and secure the manufacturing.

Design Procedures of Process Planning

Preliminary analysis of a mechanical part. Firstly, the function, importance and working condition of the part should be clarified. Secondly, the main technical demand and why it should be specified must be understood. Lastly, technical inspection of the engineering drawing must be carried out. The content of inspection includes, whether there are full and correct views, full and reasonable technical specifications, dimensions, roughness and tolerance specifications, and whether or not the structure of the part is convenient to be machined, assembled and raise the production efficiency. If the design details are considered unreasonable or wrong during inspection, suggestions to modify the design can be proposed.

Determination of production type according to production expectation.

Determination of raw material type. If the type of raw material is different, the process planning is also different, even for the same part. Therefore, before designing the operation process, a fully correct understanding to the raw material condition should be made. The determination of raw material is related with the part geometrical shape, dimensions, mechanical performance of material and production type, also related with current production conditions of the workshop.

Process route planning includes selection of mounting surface and machining method, classification of process phases, sequencing the operations, determining the degree of operation distribution, arrangement of hot working or inspection and other auxiliary operations such as cleaning, deburring, demagnetizing and chamfering.

Selection of machine tools, technical equipment (including cutting tools, fixtures and inspection tools) and auxiliary tools for each operation.

Selection of metal cutting parameters and standard operation times.

Technical economical analysis.

Form the process file.

Selection of Positioning Reference (Datum)

Those points, lines or surfaces of the part which are used to define the position of other points, lines or surfaces of the part are called the positioning reference datum.

Classification of positioning reference. Positioning reference can be classified into two categories according to application purpose.

Design reference. The points, lines or surfaces that are used by designers to determine the dimensions or relative positions of other points, lines or surfaces are called the design reference.

Process reference. The reference which is used in the process of operation or assembling is called the process reference or manufacturing reference. According to its application purpose, it can be further classified into four subcategories:

1) Operation reference. In engineering drawings, the points, lines or surfaces that are used to determine the operation dimensions or positions relative to the machined surface are called the operation reference.

2) Positioning reference. The positioning reference is defined as the points, lines or surfaces which are used to determine the locations of points, lines or surfaces that are to be machined. The process that determine the location is called positioning.

3) Inspection reference. The points, lines or surfaces which are used as a reference to inspect the locations of machined surfaces are called the inspection reference. In most cases, the design reference is used as the inspection reference.

4) Assembling reference. In the process of assembling, the points, lines or surfaces which are used as a reference to determine the locations of the parts or components relative to the product are called the assembling reference.

Unit 2　Forming of Gear Teeth

Text

There are a large number of ways of forming the teeth of gears, such as sand casting, shell molding, investment casting, permanent-mold casting, die casting and centrifugal casting. Teeth can be formed by using the powder-metallurgy process; or, by using extrusion, a single bar of aluminum may be formed and then *sliced* into gears. Gears which carry large loads in comparison with their size are usually made of steel and are cut with either form cutters or generating cutters in form cutting, the tooth space takes the exact form of the cutter. In generating, a tool having a shape different from the tooth profile is moved relative to the gear blank so as to obtain the proper tooth shape. One of the newest and most promising methods of forming teeth is called cold forming, or cold rolling, in which dies are rolled against steel blanks to form the teeth. The mechanical properties of the metal are greatly improved by the rolling process, and a high-quality generated profile is obtained at the same time.

Gear teeth may be machined by milling, shaping, or *hobbing*. They may be finished by *shaving*, *burnishing*, grinding, or *lapping*.

Milling. Gear teeth may be cut with a form milling cutter shaped to conform to the tooth space. With this method it is theoretically necessary to use a different cutter for each gear, because a gear having 25 teeth, for example, will have a different-shaped tooth space from one having, say, 24 teeth. Actually, the change in space is not too great, and it has been found that eight cutters may be used with reasonable accuracy to cut any gear in the range of 12 teeth to a *rack*. A separate set of cutters is, of course, required for each *pitch*.

Shaping. Teeth may be generated with either a pinion cutter or a rack cutter. The pinion cutter reciprocates along the vertical axis and is slowly fed into the gear blank to the required depth. When the pitch circles are tangent, both the cutter and the blank rotate slightly after each cutting stroke. Since each tooth of the cutter is a cutting tool, the teeth are all cut after the blank has completed one rotation. The sides of an involute rack tooth are straight. For this reason, a rack-generating tool provides an accurate method of cutting gear teeth. In operation, the cutter reciprocates and is first fed into the gear blank until the circles are tangent. Then, after each cutting stroke, the gear blank and cutter roll slightly on their pitch circles. When the blank and cutter have rolled a distance equal to the circular pitch, the cutter is returned to the starting point, and the process is continued until all the teeth have been cut[①].

Hobbing. The hobbing is simply a cutting tool which is shaped like a worm. The teeth have straight sides, as in a rack, but the hob axis must be turned through the lead angle in order to cut spur-gear teeth. For this reason, the teeth generated by a hob have a slightly different shape from those generated by a rack cutter. Both the hob and the blank must be rotated at the proper angular-velocity ratio. The hob is then fed slowly across the face of the blank until all the teeth have been cut.

Finishing. Gears which run at high speeds and transmit large forces may be subjected to additional *dynamic* forces if there are errors in tooth profiles②. Errors may be diminished somewhat by finishing the tooth profiles. The teeth may be finished, after cutting, by either shaving or burnishing. Several shaving machines are available which cut off a minute amount of metal, bringing the accuracy of the tooth profile within the limits of 250μm.

Burnishing, like shaving, is used with gears which have been cut but not heat-treated. In burnishing, hardened gears with slightly oversize teeth are run in mesh with the gear until the surfaces become smooth.

Grinding and lapping are used for hardened gear teeth after heat treatment. The grinding operation employs the generating principle and produces very accurate teeth. In lapping, the teeth of the gear and lap slide *axially* so that the whole surface of the teeth is *abraded* equally.

Questions

1. What are the advantages of cold roll-forming for making gears?
2. What are the three basic processes for machining gears?
3. What is the diametral pitch of a gear?
4. What is the relationship between the diametral pitch and the module of a gear?
5. What are the advantages of helical gears compared with spur gears?

New Words and Expressions

1. slice [slais] n. 片，切片 vt. 切开
2. hobbing [ˈhɔbiŋ] n. 滚齿加工
3. shaving [ˈʃeiviŋ] n. 剃，刮，削片；剃齿
4. burnish [ˈbəːniʃ] vt. 抛光，磨光 n. 光亮
5. lapping [ˈlæpiŋ] n. 研磨，磨片
6. rack [ræk] n. 齿条，机架
7. pitch [pitʃ] n. 节距，齿节
8. dynamic [daiˈnæmik] adj. 动态的，电动的 n. 动力
9. axial [ˈæksiəl] adj. 轴的，轴向的
10. abrade [əˈbreid] vt. 擦，擦破；磨，磨损
11. dedendum [diˈdendəm] n. 齿根高
12. addendum [əˈdendəm] n. 齿顶高
13. flank [flæŋk] n. 齿面 vt. 位于……的侧面
14. undercut [ˈʌndəkʌt] n. 根切，切槽
15. a large number of 许多，很多

16. in comparison with 和……比起来
17. be in mesh （齿轮）互相啮合

Notes

[1] When the blank and cutter have rolled a distance equal to the circular pitch, the cutter is returned to the starting point, and the process is continued until all the teeth have been cut.

当坯料和刀具转动的距离与齿距相等时，刀具就回到了起点，并且这个过程会持续到所有轮齿全部加工完为止。

句中 equal to 可译为“相等的，相同的”；until 为连词，可译为“直到……为止，在……以前”。

[2] Gears which run at high speeds and transmit large forces may be subjected to additional dynamic forces if there are errors in tooth profiles.

如果齿廓存在误差，高速运转和传递较大力的齿轮就可能会受到附加动态力的影响。

句中 which run at high speeds and transmit large forces 为限定性定语从句，修饰 gears；if 为连接词，引导真实条件从句，可译为“如果”。例：Work is done if we lift a weight. 可译为“如果我们举起重物，我们就做了功。”

Glossary of Terms

1. spur gear 直齿轮
2. helical gear 斜齿轮
3. bevel gear 伞齿轮
4. worm and worm gear 蜗杆和蜗轮
5. spiral gear 螺旋齿轮
6. hypoid gear 偏轴伞齿轮，准双曲面齿轮
7. herringbone gear 人字形齿轮
8. pinion 小齿轮
9. top gear 高速齿轮
10. involute gear 渐开线齿轮
11. pitch circle 节圆
12. base circle 基圆
13. circular pitch 齿距
14. diametral pitch 径节
15. dedendum circle 齿根圆
16. addendum circle 齿顶圆
17. face width 齿宽
18. gear blank 齿坯
19. gear hobbing machine 滚齿机
20. factor of safety（safety coefficient） 安全系数

21. tooth curve 齿向曲线
22. tooth error 齿形误差
23. shaving machine 剃齿机
24. burnishing machine 抛光机
25. feed rack 进给齿条
26. steering rack 转向齿条
27. disengage gear 齿轮离合器
28. tooth head 齿顶（高）

Reading Materials

Types of Gears

Spur gears have teeth parallel to the axis of rotation and are used to transmit motion from one shaft to another, parallel, shaft. Of all types, the spur gear is the simplest and, for this reason, will be used to develop the primary kinematic relationships of the tooth form.

Helical gears have teeth inclined to the axis of rotation. Helical gears can be used for the same applications as spur gears and, when so used, are not as noisy, because of the more gradual engagement of the teeth during meshing. The inclined tooth also develops thrust loads and bending couples, which are not present with spur gearing. Sometimes helical gears are used to transmit motion between non-parallel shafts.

Bevel gears, have teeth formed on conical surfaces and are used mostly for transmitting motion between intersecting shafts. Spiral bevel gears are cut so the tooth is no longer straight, but forms a circular arc. Hypoid gears are quite similar to spiral bevel gears except that the shafts are offset and non-intersecting.

The fourth basic gear type is the worm and worm gear. The direction of rotation of the worm gear, also called the worm wheel, depends upon the direction of rotation of the worm and upon whether the worm teeth are cut right-hand or left-hand. Worm-gear sets are also made so that the teeth of one or both wrap partly around the other. Such sets are called single-enveloping and double-enveloping worm-gear sets. Worm-gear sets are mostly used when the speed ratios of the two shafts are quite high, say, 3 or more.

Nomenclature of Spur-gear Teeth

The terminology of spur-gear teeth is illustrated in Figure 6-2-1. The pitch circle is a theoretical circle upon which all calculations are usually based; its diameter is the pitch diameter. The pitch circles of a pair of mating gears are tangent to each other. A pinion is the smaller of two mating gears. The larger is often called the gear.

The circular pitch p is the distance, measured on the pitch circle, from a point on one tooth to a corresponding point on an adjacent tooth. Thus the circular pitch is equal to the sum of the tooth thickness and the width of space.

The module m is the ratio of the pitch diameter to the number of teeth. The customary unit of length used is the millimeter. The module is the index of tooth size in SI (International System of Units).

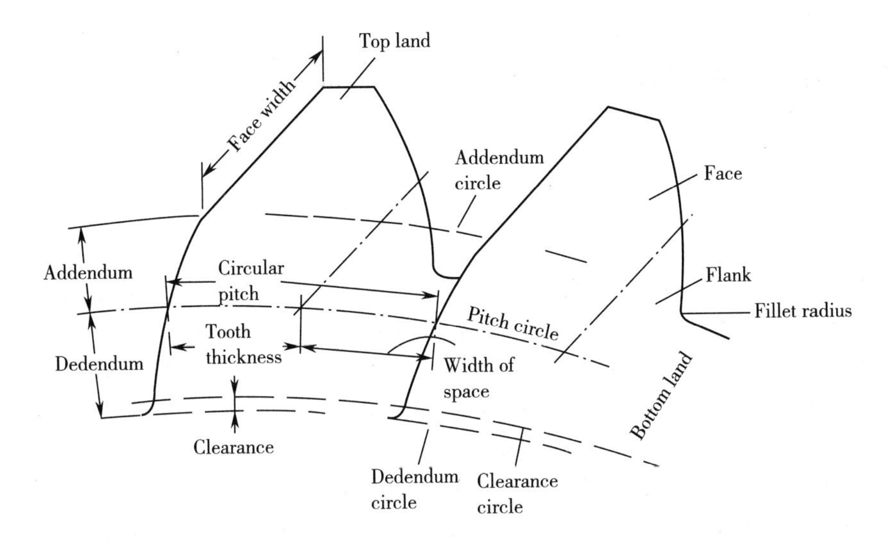

Figure 6-2-1 Nomenclature of spur-gear teeth

The diametral pitch p is the ratio of the number of teeth on the gear to the pitch diameter. Thus, it is the reciprocal of the module. Since diametral pitch is used only with U. S. units, it is expressed as teeth per inch.

The addendum a is the radial distance between the top land and the pitch circle. The dedendum b is the radial distance from the bottom land to the pitch circle. The whole depth h_t is the sum of the addendum and the dedendum.

The clearance circle is a circle that is tangent to the addendum circle of the mating gear. The clearance c is the amount by which the dedendum in a given gear exceeds the addendum of its mating gear. The backlash is the amount by which the width of a tooth space exceeds the thickness of the engaging tooth measured on the pitch circles.

Unit 3 Shaft Design

Text

In general, the shafts can be divided into three classes (shaft, axle, spindle). A shaft is a rotating member, usually of circular cross section, used to transmit power or motion. It provides the axis of rotation, or *oscillation*, of elements such us gears, *pulleys*, flywheels, cranks, *sprockets* and the like, and controls the geometry of their motion. An axle is a non-rotating member which carries no torque and is used to support rotating wheels, pulleys, and the like①. The automotive axle is not a true axle; the term is a carry-over from the horse-and-buggy era, when the wheels rotated on non-rotating members. A spindle is a short shaft. Terms such as line-shaft, head-shaft, stub-shaft, transmission shaft, countershaft, and flexible shaft are names associated with special usage.

A shaft design really begins after much *preliminary* work. The design of the machine itself will dictate that certain gears, pulleys, bearings and other elements will have at least been partially analyzed and their size and spacing tentatively determined. At this stage the design must be studied from the following points of view:

(1) Deflection and *rigidity*: (a) Bending deflection; (b) Torsional deflection; (c) *Slope* at bearings and shaft-supported elements; (d) Shear deflection due to *transverse* loading of short shafts.

(2) Stress and strength: (a) Static strength; (b) Fatigue strength; (c) Reliability.

The geometry of a shaft is generally that of a stepped cylinder. Gears, bearings and pulleys must always be accurately positioned and provision is made to accept thrust loads. The use of shaft shoulders is an excellent means of axially locating the shaft elements; these shoulders can be used to preload rolling bearings and to provide the necessary thrust reactions to the rotating elements. For these reasons our analysis will usually involve stepped shafts.

There is no *magic formula* to determine the shaft geometry for any given design situation. The best approach is that of studying existing designs to learn how problems have been solved and then of combining the best of these to solve own problem.

Many shaft-design situations include the problem of transmitting torque from one element to another on the shaft. Common torque-transfer elements are: keys, splines, *setscrews*, pins, press or shrink fits, tapered fits.

All these torque-transfer means solve the problem of securely *anchoring* the wheel or device to the shaft, but not all of them solve the problem of accurate axial location of the device②. Some of the most-used locational devices include: *cotter* and *washer*, nut and washer, sleeve, shaft shoulder, ring and groove, setscrew, split hub or tapered two-piece hub, collar and screw, pins.

Questions

1. How are shafts classified? Can you give an example to illustrate each kind of the shaft?

2. When designing the construction of a shaft, what should be considered?
3. How many methods are used to fix the elements on the shaft along the axial direction?
4. What are the most-used locational devices in shaft design?

New Words and Expressions

1. oscillation [ˌɔsiˈleiʃən] n. 摆动，振动
2. pulley [ˈpuli] n. 滑轮，带轮 vt. 装滑轮
3. sprocket [ˈsprɔkit] n. 链轮齿，星形轮
4. preliminary [priˈliminəri] n. 准备，初步 adj. 预备的
5. rigidity [riˈdʒiditi] n. 刚性，硬度，稳定性
6. slope [sləup] n. 倾斜，斜度
7. transverse [ˈtrænzvəːs] adj. 横向的 n. 横轴，交叉物
8. magic [ˈmædʒik] n. 魔力 adj. 不可思议的
9. formula [ˈfɔːmjulə] n. 公式，方案
10. setscrew [ˌsetˈskruː] n. 定位，固定螺钉
11. anchor [ˈæŋkə] n. 锚，地脚螺钉 vt. 抛锚，固定
12. cotter [ˈkɔtə] n. 销，键
13. washer [ˈwɔʃə] n. 垫圈，垫片
14. be divided into 把……分成
15. one divids into two 一分为二
16. be associated with 与……联合在一起，与……结合

Notes

[1] An axle is a non-rotating member which carries no torque and is used to support rotating wheels, pulleys, and the like.

心轴是一个不转动的构件，它不传递转矩，而是用来支承回转轮、带轮等零部件的。

句中 and the like 可译为“等等，以及诸如此类。”例：bolt, screw and the like 可译为“螺柱、螺钉等等。”

[2] All these torque-transfer means solve the problem of securely anchoring the wheel or device to the shaft, but not all of them solve the problem of accurate axial location of the device.

所有这些转矩传递方法都能够解决轴上轮子或其他装置的可靠联接问题，但不是所有的方法都能够实现装置的轴向准确定位。

句中 or 引导同义短语，意为“即，或者说”；not all 为 not 与 all 连用，表示部分否定，可译为“不是所有的”。

Glossary of Terms

1. line-shaft 总轴，动力轴
2. sprocket wheel 链轮

3. transmission shaft 传动轴
4. transmission case 变速箱
5. countershaft 副轴，中间轴
6. flexible shaft 柔性轴
7. crankshaft 曲轴
8. high-speed shaft 高速轴
9. shaft coupling 联轴器
10. nut and washer 螺母和垫圈
11. hexagon bolt 六角螺栓
12. hexagon nut 六角螺母
13. sleeve 滑套，套筒
14. shaft shoulder 轴肩
15. shaft journal 轴颈
16. bending rigidity 弯曲刚度
17. rolling-type（antifriction）bearing 滚动轴承
18. ball bearing 滚珠（球）轴承
19. roller bearing 滚柱轴承
20. radial bearing 向心（径向）轴承
21. thrust bearing 推力轴承
22. sliding（sleeve）bearing 滑动轴承
23. bearing table 轴承座
24. bearing sleeve 轴承衬，轴承套
25. tapered bearing 圆锥轴承
26. step bearing 端轴承
27. cross head cotter 十字头销
28. cotter pin 开口销
29. rigid（flexible）coupling 刚性（柔性）联轴器
30. solid（hollow）shaft 实心（空心）轴

Reading Materials

Bearing Types

Bearings are manufactured to take pure radial loads, pure thrust loads, or a combination of the two kinds of loads. The nomenclature of a ball bearing is illustrated in Figure 6-3-1, which also shows the four essential parts of a bearing. These are the outer ring, the inner ring, the balls or rolling elements and the separator. In low-priced bearings, the separator is sometimes omitted, but it has the important function of separating the elements so that rubbing contact will not occur.

In this section we include a selection from the many types of standardized bearings which are manufactured. Most bearing manufacturers provide engineering manuals and brochures containing lavish descriptions of the various types available. In the small space available here, only a meager outline of some of the most common types can be given. So you should include a survey of bearing manufacturers, literature in your studies of this section.

Some of the various types of standardized bearings which are manufactured are shown in Figure 6-3-2. The single-row deep-groove bearing will take radial load as well as some thrust load. The balls are inserted into the grooves by moving the inner ring to an eccentric position. The balls are separated after loading, and the separator is then inserted.

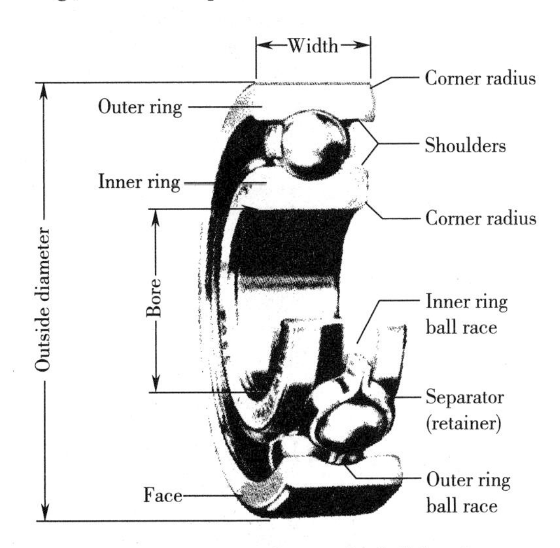

Figure 6-3-1 Nomenclature of a ball bearing

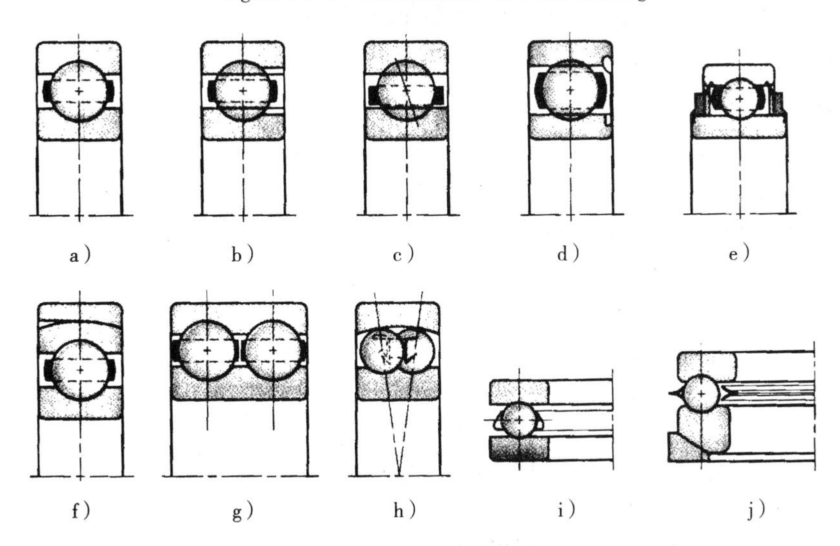

Figure 6-3-2 Various of ball bearings

a) Deep groove b) Filling notch c) Angular contact d) Shielded e) Sealed
f) External self-aligning g) Double row h) Self-aligning i) Thrust j) Self-aligning thrust

Fastenings

In a machine there are many elements, which are necessarily to connect each other, due to construction, manufacturing, fitting, transportation and service. If the two parts of the fastener are fixed, and cannot move relatively, this is called static fastening. Otherwise, it is called kinematic fastening, i. e. the two parts of the fastener can make a specific relative motion. In general, the fastening is mainly the static fastening.

All fastenings can be divided into two classes—disconnectable fastenings and permanent joints. Figure 6-3-3 shows the classification of fastenings.

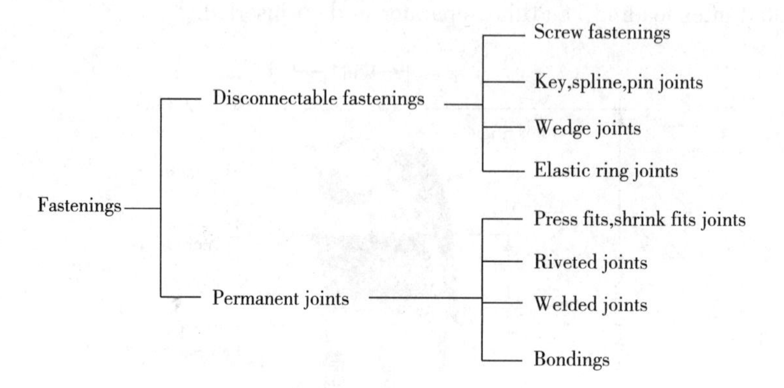

Figure 6-3-3 Classification of fastenings

Besides, according to the working principle of load transmission, the fastenings can also be divided into two classes, one is friction, and the other is non-friction. Press fits and shrink fits are the former, while the flat key fastenings are non-friction joints.

Thread Standards and Definitions

The terminology of screw threads, illustrated in Figure 6-3-4, is explained as follows: the pitch is the distance between adjacent thread forms measured parallel to the thread axis. The pitch in U. S. units is the reciprocal of the number of thread forms per inch N.

The major diameter d is the largest diameter of a screw thread.

The minor diameter d_r or d_1 is the smallest diameter of a screw thread.

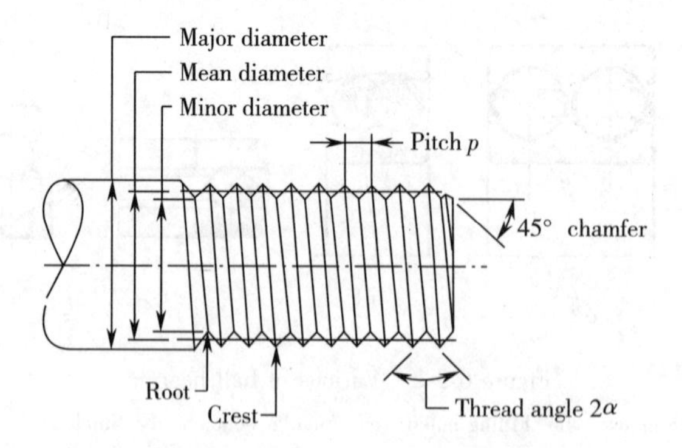

Figure 6-3-4 Terminology of screw threads

The lead l not shown, is the distance the nut moves parallel to the screw axis when the nut is given one turn. For a single thread, the lead is the same as the pitch.

A multiple-threaded product is one having two or more threads cut beside each other (imagine two or more strings wound side by side around a pencil). Standardized products such as screws, bolts, and nuts all have single thread; a double-threaded screw has a lead equal to twice the pitch, a triple-threaded screw has a lead equal to 3 times the pitch, and so on.

All threads are made according to the right-hand rule unless otherwise noted.

The American National (Unified) thread standard has been approved in this country and in Great Britain for use on all standard threaded products. The thread angle is 60° and the crests of the thread may be either flat or rounded.

Chapter 7 CAD/CAM and Numerical Control Machines

Unit 1 CAD/CAM

Text

Since the *advent* of computer technology, manufacturing professionals have wanted to automate the design process and use the *database* developed therein for automating manufacturing processes. Computer-aided design/Computer-aided manufacturing (CAD/CAM), when successfully *implemented*, should remove the "wall" that has traditionally existed between the design and manufacturing of components.

CAD/CAM means using computers in the design and manufacturing processes. Since the advent of CAD/CAM, other terms have developed:

(1) Computer graphics (CG)

(2) Computer-aided engineering (CAE)

(3) Computer-aided design and drafting (CADD)

(4) Computer-aided process planning (CAPP)

These spin-off terms all refer to specific aspects of the CAD/CAM concept[①]. CAD/CAM itself is a broader, more inclusive term. It is at the heart of automated and integrated manufacturing.

A key goal of CAD/CAM is to produce data that can be used in manufacturing a product while developing the database for the design of that product. When successfully implemented, CAD/CAM involves the sharing of a common database between the design and manufacturing of components of a company.

Interactive computer graphics (ICG) plays an important role in CAD/CAM. Through the use of ICG, designers develop a graphic image of the product being designed while storing the data that electronically make up the graphic image[②]. The graphic image can be presented in a two-dimensional (2-D), three-dimensional (3-D), or solids format. ICG images are constructed using such basic geometric characters as points, lines, circles and curves. Once created, these images can be easily edited and *manipulated* in a variety of ways including *enlargements*, reductions, rotations and movements.

An ICG system has three main components: hardware, which consists of the computer and various peripheral devices; software, which consists of the computer programs and technical manuals for the system (the popular ICG software used in CAD/CAM at present includes AutoCAD, CADKEY, Pro/E, UG, I-DEAS, CATIA, etc.); and the human designer, the most important of the three components.

A typical hardware configuration for an ICG system includes a computer, a display *terminal*, a disk drive unit for *floppy diskettes*, a hard disk; and input/output devices such as a *keyboard*, *plotter* and *printer*. These devices, along with the software, are the tools modern designers use to develop and document their designs.

Questions

1. What are the main advantages of CAD over traditional drawing methods?
2. What is a key goal of CAD/CAM?
3. What are the three main components of an ICG system?
4. What is a typical hardware configuration for an ICG system?

New Words and Expressions

1. advent [ˈædvənt] n. 出现，到来
2. database [ˈdeitəˌbeis] n. 资料库，数据库
3. implement [ˈimplimənt] n. 工具，器械 vt. 执行，完成
4. manipulate [məˈnipjuleit] vt. 熟练使用，操作，处理
5. enlargement [inˈlɑːdʒmənt] n. 放大，扩大，扩建
6. terminal [ˈtəːminl] adj. 末端的，终点的 n. 末端，端子
7. floppy [ˈflɔpi] adj. 松软的，松懈的
8. diskette [ˈdisket] n. 磁盘
9. keyboard [ˈkiːbɔːd] n. 键盘，开关板
10. plotter [ˈplɔtə] n. 绘图仪，标图员
11. printer [ˈprintə] n. 打印机，晒图机
12. hard disk 硬盘
13. spin-off 伴随的，附带的效果
14. play an important role 起重要作用
15. with the advent of 随着……的到来；随着……的出现

Notes

[1] These spin-off terms all refer to specific aspects of the CAD/CAM concept.

以上这些派生术语都是指 CAD/CAM 概念下的某些专门领域。

句中 spin-off terms 可译为“派生术语”；refer to 可译为“指”。

[2] Through the use of ICG, designers develop a graphic image of the product being designed while storing the data that electronically make up the graphic image.

通过使用交互式计算机绘图，设计者能够生成目标设计产品的图形图像，同时存储图像

的电子化数据。

句中 being designed 做定语修饰 the product，表示进行时被动意义，意为“正在设计的产品”；that electronically make up the graphic image 为定语从句，修饰 the data。

Glossary of Terms

1. application program 应用程序
2. computer graphics（CG） 计算机绘图
3. computer-aided engineering（CAE） 计算机辅助工程
4. computer-aided business（CAB） 计算机辅助经营
5. computer-aided design and drafting（CADD） 计算机辅助设计和绘图
6. computer-aided process planning（CAPP） 计算机辅助工艺规程计划
7. computer-aided industrial design（CAID） 计算机辅助工业设计
8. computer-aided architectural design（CAAD） 计算机辅助建筑设计
9. interactive computer graphics（ICG） 交互式计算机绘图
10. automatically programmed tool（APT） 自动编程工具
11. computer-aided inspection（CAI） 计算机辅助检测
12. computer-aided quality control（CAQC） 计算机辅助质量控制
13. computer-aided fixturing design 计算机辅助夹具设计
14. computer-aided tooling design 计算机辅助工具设计
15. computer-aided tolerancing analysis 计算机辅助公差分析
16. software application 计算机软件应用
17. digital circuit design 数字电路设计
18. computer programming language 计算机程序语言
19. electronic design automation（EDA） 电子设计自动化
20. central processing unit（CPU） 中央处理单元
21. computer-aided testing（CAT） 计算机辅助测试
22. operating system 操作系统
23. programmable logic controller（PLC） 可编程逻辑控制器
24. input/output（I/O） 输入/输出
25. expert system 专家系统
26. paperless design 无纸设计
27. geometric modeling 几何造型
28. scanning system 扫描系统

Reading Materials

Rationale for CAD/CAM

The rationale for CAD/CAM is similar to that used to justify any technology-based improvement in manufacturing. It grows out of a need to continually improve productivity, quality, and in turn, competitiveness. There are also other reasons why a company might make a conversion from manual processes to CAD/CAM:

Increased productivity. Productivity in the design process is increased by CAD/CAM.

Better quality. Because CAD/CAM allows designers to focus more on actual design problems and less on time-consuming, nonproductive tasks, product quality improves with CAD/CAM.

Better communication. Design documents such as drawings, parts lists, bills of material, and specifications are tools used to communicate the design to those who will manufacture it.

Common database. This is one of the most important benefits of CAD/CAM. With CAD/CAM, the data generated during the design of a product can be used in producing the product.

Reduced prototype costs. With manual design, models and prototypes of a design must be made and tested, adding to the cost of the finished product.

Faster response to customers.

CAD

CAD is the use of a wide range of computer-based tools that assist engineers, architects and other design professionals in their design activities. It is the main geometry authoring tool within the Product Lifecycle Management process and involves both software and sometimes special-purpose hardware. Current packages range from 2D vector based drafting systems to 3D parametric surface and solid design modellers.

CAD is used to design and develop products, which can be goods used by end consumers or intermediate goods used in other products. CAD is also extensively used in the design of tools and machinery used in the manufacture of components. CAD is also used in the drafting and design of all types of buildings, from small residential types (houses) to the largest commercial and industrial types (hospitals and factories).

CAD is used throughout the engineering process from conceptual design and layout, through detailed engineering and analysis of components to definition of manufacturing methods.

CAE

Computer-Aided Engineering analysis (often referred to as CAE) is the application of computer software in engineering to analyze the robustness and performance of components and assemblies. It encompasses simulation, validation and optimization of products and manufacturing tools.

Considering information technology that provides tools and techniques to help in design and manufacturing support, the tools used are considered CAE tools. In the future CAE systems will be major providers of information to help support design teams in decision making.

In regards to information networks, CAE systems are individually considered a single node on a total information network and each node may interact with other nodes on the network.

CAE systems can provide support to businesses; this is achieved by the use of reference architectures and their ability to place information views on the business process. Reference architecture is the basis of information model, especially product and manufacturing models.

CAPP

Traditional process planning is performed manually by highly experienced planners who possess an in-depth knowledge of the manufacturing processes involved and the capabilities of the shop floor facilities. Because of the experience factor involved in planning for the physical reality of the product and in the absence of standardisation of the process, conventional process planning has largely been subjective. Moreover, this activity is highly labour intensive and often becomes tedious when dealing with a large number of process plans and revisions to those plans. Rather than carrying out an exhaustive analysis and arriving at optimal values which would be time consuming, process planners often tend to play safe by using conservative values and this situation would invariably leads to non-optimal utilization of the manufacturing facilities and longer lead times. They would also not be in a position to see whether a similar component has already been planned in view of the difficulties involved in going through all the old process plans.

Computer-Aided Process Planning (CAPP) is a means to automatically develop the process plan from the geometric image of the component. The key to development of such CAPP systems is to structure the data concerning part design, manufacturing facilities and capabilities into categories and logical relationships. CAPP thus appears to fully integrate CAD and CAM.

There are two basic approaches to CAPP, i. e. variant and generative, which are briefly discussed below.

The variant approach, which is also called retrieval approach, uses a Group Technology (GT) code to select a generic process plan from the existing master process plans developed for each part family and edits to suit the requirement of the part. The variant approach is commonly implemented with a GT coding system. Here, the parts are segmented into groups based on similarity and each group has a master plan. However, this approach is impractical in situations where small batches of widely varying parts are produced. Moreover, this method fails to capture real knowledge or expertise of process planners, and there is a danger of repeating mistakes from earlier plans that were stored in the database.

In generative approach, a process plan is created from scratch for each component without human intervention. These systems are designed to automatically synthesize process information to develop a process plan for a part. These systems contain the logic to use manufacturing databases and suitable part description schemes to generate a process plan for a particular part. Most of the contemporary CAPP systems being developed are generative in nature. Generative approach eliminates the disadvantages of the variant approach and bridges the gap between CAD and CAM.

Unit 2 NC Machines

Text

Principles of NC Machines

The basic elements and operation of a typical NC machine are shown in Figure 7-2-1.

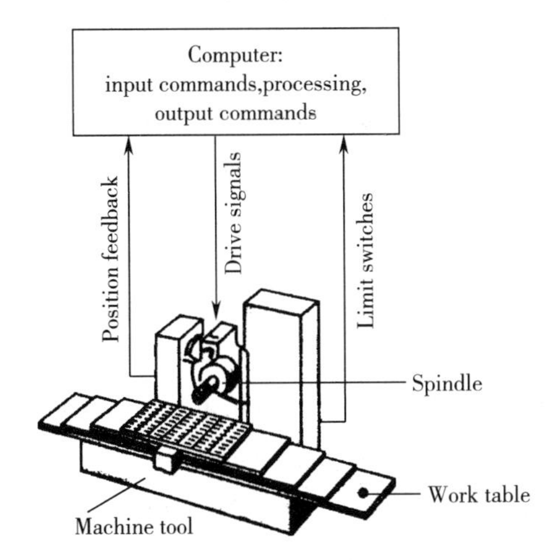

Figure 7-2-1 Schematic illustration of the major components for position control on a NC machine tool

The functional elements in numerical control and the components are as follows:

Data input: The numerical information is read and stored in the tape reader and in computer memory.

Data processing: The *programs* are read into the machine control unit for processing.

Data output: This information is translated into *commands* (typically pulsed commands) to the *servomotor*. The servomotor then moves the table (on which the workpiece is mounted) to specific positions, through linear or rotary movements, by means of stepping motors, leadscrews and other similar devices①.

Types of control circuits. A NC machine can be controlled through two types of circuits: open-loop and closed-loop. In the open-loop system (Figure 7-2-2a), the signals are sent to the servomotor by the controller, but the movements and final positions of the work table are not checked for accuracy. The closed-loop system (Figure 7-2-2b) is equipped with various *transducers*, *sensors* and counters that measure accurately the position of the work table. Through *feedback* control, the position of the work table is compared against the signal. Table movements *terminate* when the proper coordinates are reached. The closed-loop system is more complicated and more expensive than the open-loop system.

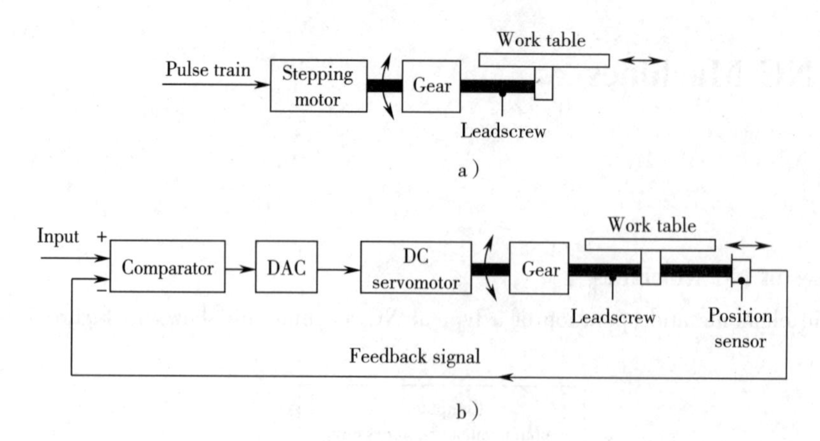

Figure 7-2-2 Schematic illustration of the components of an open-loop and a closed-loop control system for a NC machine

a) Open-loop system b) Closed-loop system

Computer Numerical Control

Two types of computerized systems were developed: direct numerical control and computer numerical control.

In **direct numerical control** (**DNC**), as originally conceived and developed in the 1960s, several machines are directly controlled, step by step, by a central *mainframe* computer. In this system, the operator has access to the central computer through a remote terminal. In this way, the handling of tapes and the need for a separate computer on each machine are eliminated. With DNC, the status of all machines in a manufacturing facility can be monitored and assessed from the central computer. However, DNC has a crucial disadvantage: If the computer shuts down, all the machines become *inoperative*.

Computer numerical control (**CNC**) is a system in which a control microcomputer is an integral part of a machine or a piece of equipment. The part program may be prepared at a remote site by the programmer, and it may incorporate information obtained from drafting software packages and from machining simulations, in order to ensure that the part program is bug free[②]. The machine operator can, however, easily and manually program on board computers. The operator can *modify* the programs directly, prepare programs for different parts, and store the programs.

Some advantages of CNC over conventional NC systems are the following: increased flexibility, greater accuracy and more versatility.

Questions

1. Describe the principles of numerical control of machines.
2. What is the difference between an open-loop and a closed-loop NC system?
3. What is direct numerical control?
4. What is computer numerical control?

New Words and Expressions

1. program [ˈprəugræm] n. 程序，方案 vt. 设计程序 vi. 编程序
2. command [kəˈmɑːnd] n. 命令，指令 vt. 命令，管理 vi. 控制，指挥
3. servomotor [ˈsəːvəuˌməutə] n. 伺服电动机，继动器
4. transducer [trænsˈdjuːsə] n. 转换器，变频器，发送器
5. sensor [ˈsensə] n. 传感器，敏感元件
6. feedback [ˈfiːdbæk] n. 反馈，回传
7. terminate [ˈtəːmineit] v. 终止，端接 adj. 终止的，有限的
8. mainframe [ˈmeinfreim] n. 总配线架，主机架
9. inoperative [inˈɔpərətiv] adj. 不起作用的，不工作的
10. modify [ˈmɔdifai] vt. 修改，更改，改进
11. translate into 把……转化成，把……翻译成
12. be equipped with 用……装备
13. step by step 一步一步，逐步

Notes

[1] The servomotor then moves the table (on which the workpiece is mounted) to specific positions, through linear or rotary movements, by means of stepping motors, leadscrews and other similar devices.

借助于步进电动机、丝杠和其他类似装置，伺服电动机驱动工作台（固定工件的装置）通过线性运动或旋转运动到达指定位置。

句中 through linear or rotary movements 为介词短语，可译为“通过线性运动或旋转运动”；by means of 为介词短语，可译为“借助于”。

[2] The part program may be prepared at a remote site by the programmer, and it may incorporate information obtained from drafting software packages and from machining simulations, in order to ensure that the part program is bug free.

程序员可以进行远程零件编程，并通过整合从绘图软件包和加工仿真获取的信息确保零件程序没有问题。

句中 in order to 为表示目的的不定式短语，翻译成汉语时也可放在主句前面。

Glossary of Terms

1. data input 数据输入
2. data processing 数据处理
3. data output 数据输出
4. open-loop control 开环控制
5. closed-loop control 闭环控制
6. semi-closed loop control 半闭环控制

7. control system 控制系统
8. numerical control (NC) 数控
9. analog control 模拟控制
10. computer numerical control (CNC) 计算机数字控制
11. direct numerical control (DNC) 直接数字控制
12. drafting software package 绘图软件包
13. control loops unit (CLU) 控制循环单元
14. servo system 伺服系统
15. feedback unit 反馈单元
16. machine control unit (MCU) 机床控制单元
17. machine coordinate origin 机床坐标原点
18. workpiece coordinate system 工件坐标系
19. machine zero 机床零点
20. machine tool reference position 机床参考位置
21. absolute dimension 绝对尺寸
22. absolute coordinate 绝对坐标
23. incremental dimension 增量尺寸
24. incremental coordinate 增量坐标
25. line interpolation 直线插补
26. circular interpolation 圆弧插补
27. clockwise arc 顺时针圆弧
28. counterclockwise arc 逆时针圆弧
29. manual part programming 手工零件编程
30. computer part programming 计算机零件编程
31. tool path 刀具轨迹
32. tool offset 刀具偏置
33. tool mark 走刀痕迹，切削痕迹

Reading Materials

Types of Numerical Control Systems

There are two basic types of control systems in numerical control (NC): point-to-point and contouring.

In a point-to-point system, also called positioning, each axis of the machine is driven separately by leadscrews and, depending on the type of operation, at different velocities. The machine moves initially at maximum velocity in order to reduce nonproductive time, but decelerates as the tool approaches its numerically defined position. Thus, in an operation such as drilling or punching a hole, the positioning and cutting take place sequentially. After the hole is drilled or punched, the tool retracts upward and moves rapidly to another position, and the operation is repeated. The path followed from one position to another is important in only one respect: it must be chosen to minimize the time of travel, for better efficiency. Point-to-point systems are used mainly in drilling, punching and straight milling operations.

In a contouring system (also known as a continuous path system), the positioning and the operations are both performed along controlled paths but at different velocities. Because the tool acts as it travels along a prescribed path, accurate control and synchronization of velocities and movements are important. The contouring system is typically used on lathes, milling machines, grinders, welding machinery and machining centers.

Advantages and Limitations of NC

NC has the following advantages over conventional methods of machine control:

Flexibility of operation is improved, as well as the ability to produce complex shapes with good dimensional accuracy, good repeatability, reduced scrap loss, high production rates, high productivity and high product quality.

Tooling costs are reduced, because templates and other fixtures are not required.

Machine adjustments are easy to make with microcomputers and digital readouts.

More operations can be performed with each setup, and the lead time for setup and machining required is less, as compared to conventional methods. Furthermore, design changes are facilitated, and inventory is reduced.

Programs can be prepared rapidly, and they can be recalled at any time, by utilizing microprocessors. Less paperwork is involved.

Faster prototype production is possible.

Operator skill required is less than that for a qualified machinist, and the operator has more time to attend to other tasks in the work area.

The major limitations of NC are the relatively high initial cost of the equipment, the need and cost for programming and computer time, and the special maintenance that requires trained personnel. Because NC machines are complex systems, breakdowns can be costly, so preventive maintenance is essential. However, these limitations are often easily outweighed by the overall economic advantages of NC.

The CNC Process

In principal, the process of CNC manufacturing is the same as conventional manufacturing methods. Conventionally, shop drawings are generated by design engineers, who pass them to machinists. The machinists then read the drawings and methodically calculate tool paths, cutter speeds, feeds, machining time, and the like. In CNC programming, the machinist still has sole responsibility for the machine's operation. However, control is no longer exercised by manually turning the axis handwheels but through programming the use of the controller. This is not to say that most proficient machinists will be computer programmers. Early CNC machines required manual input of G and M codes, but today a computer specialist is no longer needed for this task. Note the following steps in CNC processing for both conventional and computer-aided methods.

Flow of Conventional CNC Processing

1. Develop or obtain the part drawing.

2. Decide which machine (s) will perform the operations needed to produce the part.

3. Decide on the machining sequence and decide on cutter-path directions.

4. Choose the tooling required.

5. Do the required math calculations for the program coordinates.

6. Calculate the spindle speeds and feed rates required for the tooling and part material.

7. Write the CNC program.

8. Prepare setup sheets and tool lists (these will also be used for manufacturing operators).

9. Verify and edit the program, using either a virtual machine simulator such as CNCez or on the actual machine tool, creating a prototype.

10. Verify and edit the program on the actual machine and make changes to it if necessary.

11. Run the program and produce the final part.

Flow of Computer-Aided CNC Processing

1. Develop or obtain the three-dimensional geometric model of the part, using CAD.

2. Decide which machining operations and cutter-path directions are required to produce the part (sometimes computer assisted or from engineering drawings and specifications).

3. Choose the tooling to be used (sometimes computer assisted).

4. Run a CAM software program to generate the CNC part program, including the setup sheets and list of tools.

5. Verify and edit the program, using a virtual machine simulator such as CNCez.

6. Download the part program (s) to the appropriate machine (s) over the network and machine the prototype. (Sometimes multiple machines will be used to fabricate a part.)

7. Verify the program (s) on the actual machine (s) and edit them if necessary.

8. Run the program and produce the part. If in a production environment, the production process can begin.

Unit 3 Machining Centers

Text

The *trend* in NC machines is the type of machine in which a number of operations can be done in a single setup, which is called a "machining center"[①]. A machining center (Figure 7-3-1 a, b) is built to do a number of operations such as milling, drilling, boring, facing, spotting, counterboring, threading and tapping in a single setup on several surfaces of a workpiece. One relatively unskilled operator can often attend two machining centers and sometimes more. A numerically controlled machining center must have the following characteristics:

At least three linear axes of positioning control: X, Y, and Z.

Index table to permit *access* to more than one side of the workpiece automatically.

Automatic tool changer, a tool *magazine* to hold a large selection of tools, and a tool change arm that selects and places the tools into the machine spindle.

Ability to select spindle speeds and feed rates, and to bring tools to the workpiece automatically on command.

Miscellaneous functions such as coolants off/on, part clamping, program start *alerts*, peck drilling, and other automatic functions.

The cutting tool is automatically selected from the storage magazine. The number of tools in the magazine has no theoretical limit but is typically 24, 48, or 60, etc.

Machining centers have X, Y, and Z motions within a rectangular coordinate *description* of locations. However, more complex machining centers are capable of many other axis movements. These motions include tilt and swivel of the spindle head and column as well as rotation and even tilt of the work-holding table.

The machining center offers the highest potential for NC utilization and *versatility*. Its multi-axis capability has made it practical in many cases to combine several components into a single part that is machined from casting, forging, plate, or bar stock. Design configurations that were previously impossible can be achieved if enough axes of motion can be simultaneously controlled to develop the configuration.

Most often *preset* tooling for machining centers comes from the *toolroom*, but sometimes machine operators have both the time and ability to make up the preset tools needed for their location. Many shops maintain a file of tool-setting drawings for use by both programmers and toolroom employees. These drawings describe the cutter, identify it by a tool number, and establish a setting length.

Tooling for machining centers has wide *ramifications*. For example, a hole drilled through the wall of a casting has no critical depth setting, although a bore may have critical dimensions for both depth and diameter, even to the extent that the hole is checked and the adjustable boring head reset based on the actual metal removal[②].

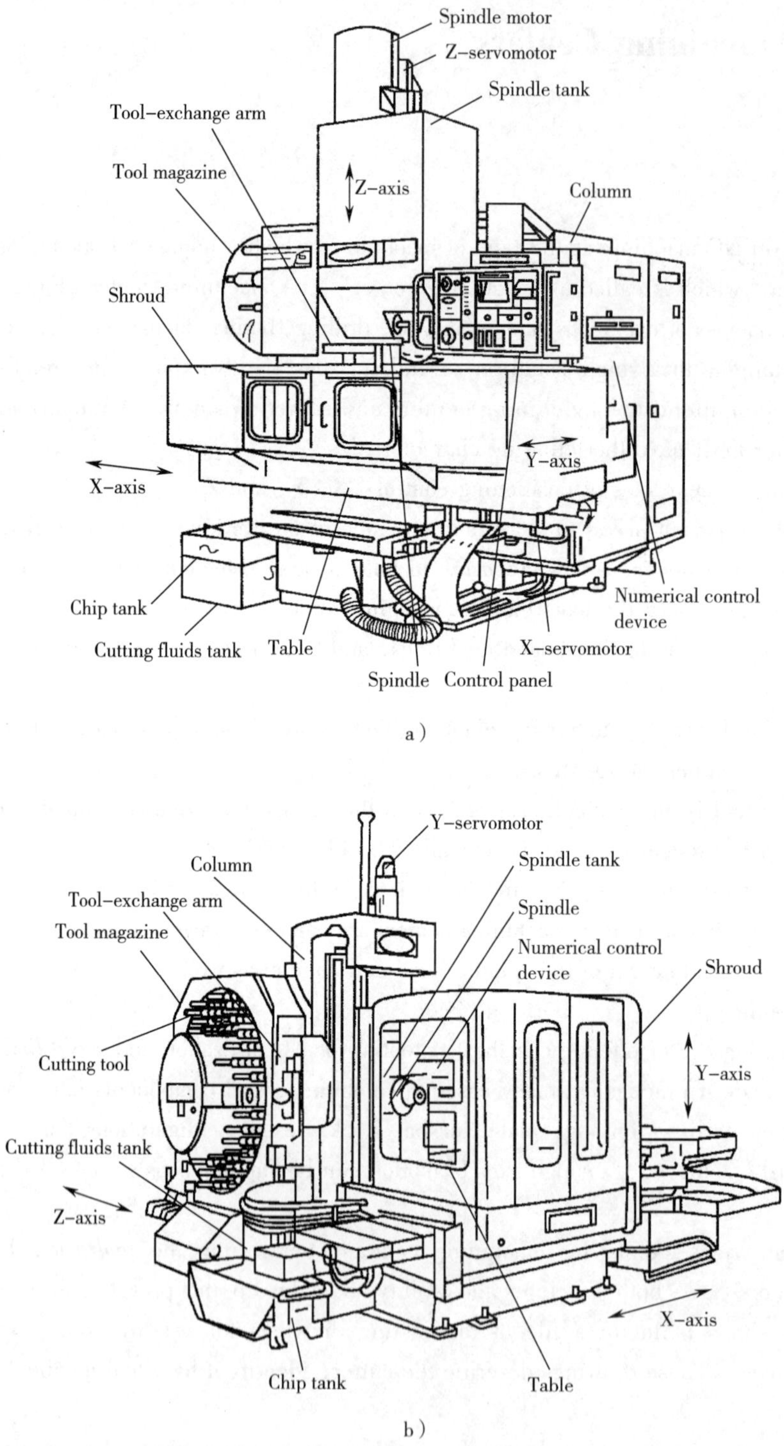

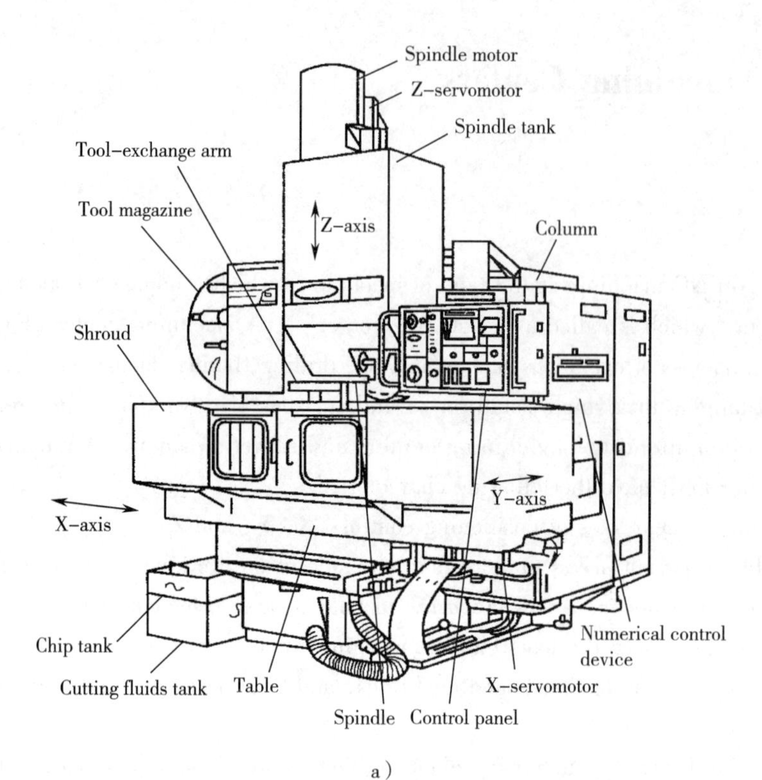

a)

b)

Figure 7-3-1 NC machining centers

a) Vertical-spindle machining center b) Horizontal-spindle machining center

Questions

1. What is the machining center?
2. What are the main operations of the machining center?
3. What are the characteristics of the machining center?

New Words and Expressions

1. trend [trend] n. & vi. 趋势，倾向
2. access ['ækses] n. 接近，进入；存取，选取数据；增长
3. magazine [ˌmægə'ziːn] n. 箱，盒；仓库；期刊，杂志
4. miscellaneous [ˌmisi'leinjəs] adj. 多方面的，有各种特点的
5. alert [ə'ləːt] adj. 灵活的，精细的 n. & vt. 警报，警惕
6. description [dis'kripʃən] n. 描写，叙述，形容；说明书，绘图
7. versatility [ˌvəːsə'tiliti] n. 多功能性，多面性，变化性
8. preset [ˌpriː'set] vt. 预调，预装，预置
9. toolroom ['tuːlruːm] n. 工具室，工具间
10. ramification [ˌræmifi'keiʃən] n. 分支，细节，结果
11. be capable of 能够，可以
12. base on 把……基于；把……建立在……上
13. beyond description 难以形容
14. give (make) a description of 描述，说明

Notes

[1] The trend in NC machines is the type of machine in which a number of operations can be done in a single setup, which is called a "machining center".

数控机床的发展趋势是加工中心，在加工中心中单次安装即可完成多道工序。

句中 in which a number of operations can be done in a single setup 为限定性定语从句，修饰 the type of machine；第二个 which 引导一个非限定性定语从句；句中 a number of 意为“许多”。

[2] For example, a hole drilled through the wall of a casting has no critical depth setting, although a bore may have critical dimensions for both depth and diameter, even to the extent that the hole is checked and the adjustable boring head reset based on the actual metal removal.

例如，穿透铸件侧壁的钻孔没有临界深度的要求，但是镗孔的孔深和直径都有临界尺寸要求，甚至要进行孔（深）的检查，并根据实际材料去除量来调整镗头。

句中 to the extent that 意为“到这样的程度，以至于……；结果……；就……来说”。

Glossary of Terms

1. vertical-spindle machining center 立式加工中心
2. horizontal-spindle machining center 卧式加工中心

3. universal machining center 万能加工中心
4. tool-changing time 换刀时间
5. tool changer 换刀机构
6. tool magazine (storage) 刀库
7. tool-exchange arm 换刀机械手
8. automatic tool changer 自动换刀装置
9. computer numerical controlled turning center 计算机数控车削加工中心
10. computer numerical control panel 计算机数控面板
11. feed rate 进给率
12. magazine feed 自动传输带（送料带）
13. magazine attachment （机床）送料装置
14. traveling column 移动立柱
15. fixed column 固定立柱
16. net shape 净成形
17. near-net shape 接近最终形状
18. preset control 程序控制
19. preset sequence 给定程序
20. memory access 存取器
21. random access 随机存取
22. tooling cost 加工成本
23. tooling quality 切削性能
24. tooling zone 刀具调整区域
25. 5-axis machine 五轴联动机床
26. preparatory function 准备功能（G 功能）
27. spindle speed function 主轴转速功能（S 功能）
28. tool function 刀具功能（T 功能）
29. miscellaneous function 辅助功能（M 功能）
30. feed function 进给功能（F 功能）

Reading Materials

Types of Machining Centers

Machining centers, just like turning centers, are classified as either vertical or horizontal. Vertical machining centers continue to be widely accepted and used, primarily for flat parts and where three-axis machining is required on a single part face such as in mold and die work. The tool magazine is on the left of the figure, and all operations and movements are directed and modified through the computer-control panel shown on the right. Because the thrust forces in vertical machining are directed downward, such machines have high stiffness and produce parts with good dimensional accuracy.

Horizontal machining centers are also widely accepted and used, particularly with large boxy, and heavy parts, because they lend themselves to easy and accessible pallet shuttle transfer when used in a cell or FMS application.

Universal machining centers are more recent developments and are equipped with both vertical and horizontal spindles. They have a variety of features and are capable of machining all surfaces of a workpiece.

Characteristics of Machining Centers

The major characteristics of machining centers are summarized here:

(1) Machining centers are capable of handling a wide variety of part sizes and shapes efficiently, economically, repetitively, and with high dimensional accuracy with tolerances in the order of ± 0.0025mm.

(2) These machines are versatile and capable of quick changeover from one type of product to another.

(3) The time required for loading and unloading workpieces, changing tools, gaging of the part, and troubleshooting is reduced. Therefore productivity is improved, thus reducing labor requirements (particularly skilled labor) and minimizing production costs.

(4) These machines are equipped with tool-condition monitoring devices for the detection of tool breakage and wear as well as probes for tool-wear compensation and for tool positioning.

(5) In-process and post-process gaging and inspection of machined workpieces are now features of machining centers.

(6) These machines are relatively compact and highly automated and have advanced control systems, so one operator can attend to two or more machining centers at the same time, thus reducing labor costs.

Selection and Innovation of Machining Centers

The selection of the type and size of machining centers depends on several factors, especially:

(1) Type of products, their sizes, and their shape complexities.

(2) Type of machining operations to be performed, and the type and number of cutting tools required.

(3) Dimensional accuracy required.

(4) Production rate required.

Machining center innovations and developments have brought about the following improvements:

(1) Improved flexibility and reliability.

(2) Increased feeds, speeds, and overall machine construction and rigidity.

(3) Reduced loading, tool-changing and other non-cutting time.

(4) Greater MCU (machine control unit) capability and compatibility with systems.

(5) Reduced operator involvement.

(6) Improved safety features and less noise.

Chapter 8 Nontraditional Manufacturing

Unit 1 Nontraditional Material Removal Processes

Text

Nontraditional/unconventional machining is the general title given to the advanced process as for material removal techniques using chemical, physical (electricity, sound, light, heat, magnetism), electrochemical and other means (not including mechanical force) . Compared with the *traditional* processes, they have been developed largely because of the need to solve some problems, which could not be adequately solved before①. The high development and the great progress in this field show the brightest prospect in application.

Why do we need these alternative processes? The traditional machining processes are not satisfactory, economical, or even possible for the following reasons:

(1) The hardness, strength and melting point of the material is very high (typically the hardness is above 400HBS), or the material is too brittle.

(2) The workpiece is too *flexible*, *slender*, *fragile*, or delicate to withstand the cutting or grinding force, or the parts are difficult to clamp in workholding devices. The workpiece may be damaged by ordinary clamping or machining.

(3) The geometrical shape of the part is complex, including such features as internal and external profiles or small-diameter holes in fuel-injection nozzles.

(4) Surface finish quality and dimensional tolerance requirements are more *rigorous* than those obtained by other processes.

(5) Temperature rise and residual stresses in the workpiece are not desirable or acceptable.

The requirements led to the development of nontraditional machining. When selected, combined and applied properly, advanced machining processes, which give a broad possibility for new structures, materials and quality, offer major significant technical and economic advantages over traditional machining methods②.

The nontraditional machining methods also have some disadvantages:

(1) Parts must be designed for the appropriate process, for example, no sharp corners, deep or narrow cavities.

(2) Parts are often uneconomical unless large quantities are manufactured.

(3) Complex and/or dangerous processes must be computer numerical controlled (CNC) .

There are several technologies available now. The main processes amongst them are chemical machining (CM), electrolytic or electrochemical machining (ECM), electrochemical grinding (ECG), electrical discharge machining (EDM), ultrasonic machining (USM), laser-beam machining (LBM), water-jet machining (WJM), electro-beam machining (EBM), ion-beam machining (IBM), plasma-arc machining (PAM), abrasive water-jet machining (AWJM), abrasive jet machining (AJM), and so on.

Questions

1. What is the nontraditional machining method?
2. Why are these nontraditional machining methods needed?
3. What are the disadvantages of the nontraditional machining methods?
4. What are the main processes amongst these nontraditional machining methods?

New words and Expressions

1. nontraditional [ˌnɔntrəˈdiʃənl] adj. 非传统的，特种的
2. unconventional [ˌʌnkənˈvenʃənl] adj. 非惯例的，非常规的
3. traditional [trəˈdiʃənl] adj. 传统的，惯例的
4. flexible [ˈfleksəbl] adj. 挠性的，韧性的，灵活的
5. slender [ˈslendə] adj. 细长的，不足的
6. fragile [ˈfrædʒail] adj. 易碎的，脆的
7. delicate [ˈdelikit] adj. 精致的，灵敏的
8. rigorous [ˈrigərəs] adj. 严格的，精确的
9. melting point 熔点
10. compare with 相比，比得上
11. (be) compared to 与……相比，与……相似
12. (be) combined into 化合成

Notes

[1] Compared with the traditional processes, they have been developed largely because of the need to solve some problems, which could not be adequately solved before.

由于需要解决一些（加工过程中的）问题，而这些问题在之前是难以充分解决的，与传统的加工工艺相比，非传统（特种）加工工艺已经有了更大的发展。

句中 they 代表非传统（特种）加工工艺（nontraditional/unconventional machining processes）; because of 为复合介词，可译为“由于，因为”; which 引导非限定性定语从句，修饰 some problems。

[2] When selected, combined and applied properly, advanced machining processes, which give a broad possibility for new structures, materials and quality, offer major significant technical and economic advantages over traditional machining methods.

经过适当的选择、组合和应用，先进加工工艺为（满足）新结构、新材料和高质量的要求提供了更为广泛的可能性，其主要技术和经济支持也优于传统加工方法。

句中 which 引导非限定性定语从句，且在从句中做主语，指代 advanced machining processes。

Glossary of Terms

1. non-traditional machine tool 特种加工机床
2. chemical machining（CM） 化学加工
3. electrolytic or electrochemical machining（ECM） 电化学加工
4. electrochemical grinding（ECG） 电化学磨削
5. electrical discharge machining（EDM） 电火花加工
6. ultrasonic machining（USM） 超声波加工
7. laser beam machining（LBM） 激光束加工
8. water jet machining（WJM） 水射流加工
9. electro beam machining（EBM） 电子束加工
10. ion beam machining（IBM） 离子束加工
11. plasma arc machining（PAM） 等离子体电弧加工
12. abrasive water-jet machining（AWJM） 磨料水射流加工
13. abrasive jet machining（AJM） 磨料喷射加工
14. pulsed electrochemical machining（PECM） 脉冲电化学加工
15. nontraditional machining process 特种（非传统）加工工艺
16. unconventional machining process 非常规加工工艺
17. traditional machining process 传统的加工工艺
18. physical-chemical etching 物理化学腐蚀
19. reactive plasma etching 反应等离子体刻蚀
20. photoetching 光电腐蚀（光刻）
21. residual stress 残余应力
22. engraving machine 雕刻机
23. dielectric fluids 电解液
24. superfinishing machine 超精加工机床
25. electrolytic machine tool 电解加工机床
26. electro-discharge machine tool 放电加工机床
27. photo chemical machining（PCM） 光化学加工

Reading Materials

Chemical Machining

It is well known that chemicals attack metals and etch them. Based on the observation of this phenomenon，chemical machining（CM）was developed. Small amounts of material can be removed from the surface by controlling chemical action. This process is carried out by chemical dissolution, using reagents or etchants, such as acids and alkaline solutions.

Chemical machining is the oldest of the nontraditional machining processes, and has been used to engrave metals and hard stones, in deburring, and more recently in the production of printed-circuit boards and microprocessor chips. Chemical milling, chemical blanking and photochemical blanking are the three main approaches in chemical machining.

Chemical milling. Chemical milling is the name given to a patented process for removing large amounts of stock by etching selected areas of complex workpieces. It was developed in the aircraft industry as one means of fabricating lightweight parts of large areas and thin sections but has been receiving attention on other industries. Chemical milling entails four steps: cleaning, masking, etching and demasking. It can be done on all kinds of parts: rolled sections, forgings, castings and preformed pieces. It is not limited by shape, direction of cut, or cutter, and different sizes and shapes of cuts can be made at one time.

Chemical blanking. Chemical blanking is a modification of chemical milling. Chemical blanking is similar to the blanking of sheet metal with the exception that material is removed by chemical dissolution rather than by shearing. It is used to produce features, which penetrate through the thickness of material. Material is removed, usually from flat thin sheet by photographic techniques. Therefore, it is sometimes called photochemical machining. In addition, when the process is used for etching, it is called photoetching. Typical applications for chemical blanking are the burr-free etching of printed-circuit boards, decorative panels, thin sheet-metal stampings, as well as the production of complex or small shapes. The process is also effective for blanking fragile workpieces and materials.

Ion-Beam Machining

Ion-beam machining (IBM) works on the principle similar to the one on which the electron-beam machining does, which is that workpiece atoms are removed by the transfer of momentum from the impinging ions as they strike the workpiece surface. However, the mechanism of material removal is quite different from electron-beam machining. The process takes place in a vacuum chamber, with charged atoms (ions) fired from an ion source towards a target (the workpiece) by means of an accelerating voltage. Plasma beams are used to rapidly cut nonferrous materials (usually sheet and plate). Temperatures reach almost 10000°C and the process is automated, as these temperatures are too high for safe manual use. The surface finish is generally poor and severe hardening of the surface can occur on steels, creating problems if further machining is required.

IBM can be used for various purposes. For example, smoothing, ion-beam texturing, ion-beam cleaning, shaping, polishing, thinning and milling are all reported. The use of IBM for smoothing of laser mirrors and for modifying the thickness of thin films and membranes without affecting surface finish is reported by Jolly, Clampitt and Reader (1983).

Unit 2 Electrochemical Machines

Text

The principle *underlying electrochemical* machining (ECM) rests on an exchange of charge and material between a positively charged anodic workpiece and a negatively charged *cathode* tool in an *electrolyte*. In such conditions, the *anode* dissolves whilst the cathode (tool) is not affected. The volume of metal removal may be calculated according to Faraday's Law:

$$m = CIt$$

Where C— a constant dependent on work material;

I— the current flowing between the tool and the workpiece;

t— the time of *erosion*.

The current is dependent on the *gap* between the tool and the workpiece, the area of erosion and the conductivity of the electrolyte, as well as the supply voltage. The working gap maintained between the *electrode* and workpiece allows machining to take place without physical contact①. The electrolyte (e. g. NaCl or $NaNO_3$ in water solution) is pumped into the working gap and also serves as a *coolant* which is necessary due to the high energy density. The material which has been eroded from the work forms a *sludge*, and must be separated from the electrolyte by *filter* or centrifuges.

Electrochemical Die-sinking Machines. Figure 8-2-1 presents, in *schematic* form, the main components of an electrochemical die-sinking machine. A feeding device advances the tool towards the workpiece in accordance with the rate of metal removal. When producing internal forms, a design problem arises in relation to the shape and size of the tool. The gap is not constant, but is a function of the state of the surface to be eroded and the rate of tool advance. If, for example, a *cylindrical* bore is to be sunk, a simple cylindrical tool (as shown in Figure 8-2-2) is not suitable. This would result in a constantly increasing gap size and the current density would decrease in proportion (Figure 8-2-2 left). With a tool suitably insulated on its sides (Figure 8-2-2 right), the *offending* excessive erosion of cylindrical sides will be suppressed.

Owing to the small gap sizes which are used in electrochemical machining, high electrolyte pressures (> 20bar) are necessary so that there is an adequate flow for effective cooling and removal of the eroded material. In an erosion area of 10, 000mm^2, forces in excess of 20kN may be experienced, with which the tool-feeding system and the machine structure must be able to cope. As the electrolyte consists of a corrosive salt solution, all machine components likely to come into contact with it must be corrosion-proof. An important *auxiliary* installation is a short circuit cut-out, which immediately stops the supply of further electrical energy in the event of inadequate clearance of the eroded material and insufficient gap between tool and work②.

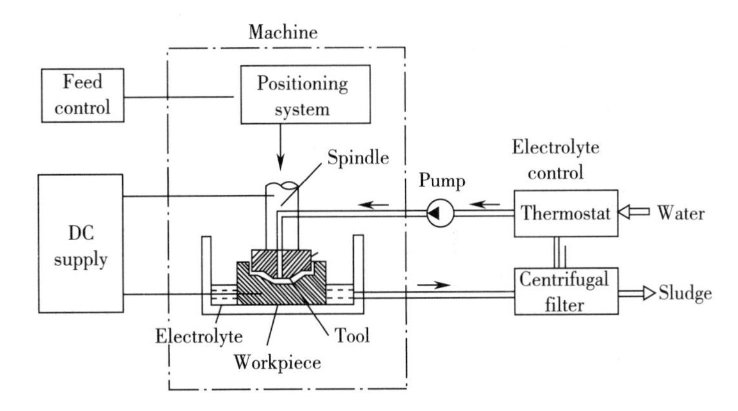

Figure 8-2-1 The main components of an electrochemical die-sinking machine

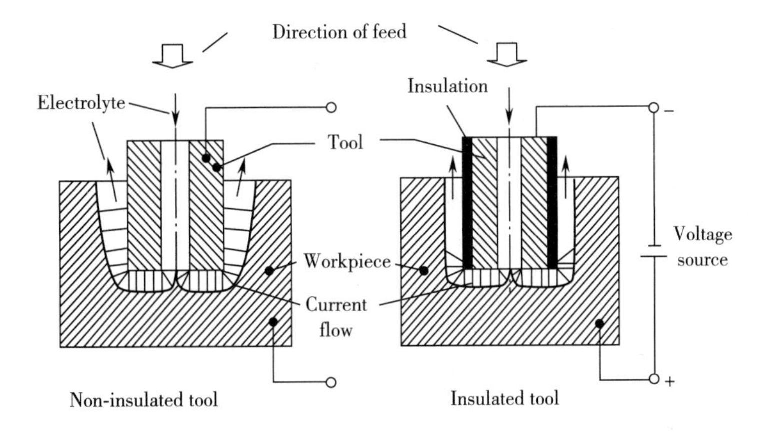

Figure 8-2-2 Cavity development during electrochemical machining with different tool electrodes

Advantages. The tool does not wear.

There are no thermal or mechanical stresses on the workpiece.

Faster stock removal and better surface finish can be obtained.

Disadvantages. The basic cost of the equipment is several times that of EDM (Electrical Discharge Machining).

Rigid fixturing is required to withstand the high electrolyte flow rates.

The tool is more difficult to make than that for EDM since it must be insulated to maintain the correct conductive paths to the workpiece.

The most common electrolyte, sodium chloride, is corrosive to the equipment, tooling and workpiece.

Questions

1. What is Faraday's Law?
2. Describe the principle of electrochemical machining.
3. What are the elements influencing the current in Faraday's Law?

New Words and Expressions

1. underlying [ˌʌndəˈlaiiŋ] adj. 在下的，基础的，潜在的
2. electrochemical [iˈlektrəuˈkemikəl] adj. 电化学的
3. cathode [ˈkæθəud] n. 阴极，负极
4. electrolyte [iˈlektrəulait] n. 电解质，电解液，电离质
5. anode [ˈænəud] n. 阳极，正极
6. erosion [iˈrəuʒən] n. 腐蚀，侵蚀，磨蚀
7. gap [gæp] n. 裂口，间隙，火花隙
8. electrode [iˈlektrəud] n. 电极
9. coolant [ˈkuːlənt] n. 冷却液，切削液
10. sludge [slʌdʒ] n. 淤泥，泥浆
11. filter [ˈfiltə] n. 过滤器，滤纸 vt. 过滤
12. schematic [skiˈmætik] adj. 图解的，按照图式的
13. cylindrical [siˈlindrikəl] adj. 圆柱体的，圆柱形的
14. offend [əˈfend] vt. 冒犯，使不愉快 vi. 违反
15. auxiliary [ɔːgˈziljəri] adj. 辅助的，附属的 n. 辅助者，附件
16. summarize [ˈsʌməraiz] vt. 概括，概述，总结
17. due to 由于，应归于
18. take place 发生
19. in relation to 关于，涉及，与……有关
20. cut-out 切断，去掉
21. short circuit 短路

Notes

[1] The working gap maintained between the electrode and work allows machining to take place without physical contact.

电极和工件之间保持的加工间隙，既避免了物理接触，又能完成加工。

单个分词 working 做 gap 的前置定语，分词短语 maintained between the electrode and work 做 gap 的后置定语。注意：当分词做定语时，一般情况下，单个分词位于被修饰词前，而分词短语则一定要位于被修饰词之后。

[2] An important auxiliary installation is a short circuit cut-out, which immediately stops the supply of further electrical energy in the event of inadequate clearance of the eroded material and insufficient gap between tool and work.

如果被腐蚀材料清理不充分或工件和刀具之间的间隙不足，机器中的重要辅助装置——短路切断器将立刻停止电能的进一步供给。

句中由 which 引导非限定性定语从句，修饰 a short circuit cut-out；从句中的 in the event of 可译为“如果……发生，万一”。

Glossary of Terms

1. underlying metal 底层金属
2. electroforming 电成形，电铸
3. perforated electrode 多孔电极
4. electrochemical machining 电化学加工
5. form of electro-machining 电加工成形面
6. electric machining 电加工
7. salt bath electrode furnace 电极盐浴炉
8. electrolytic forming 电解成形
9. electrolytic forming machine 电解型腔加工机床
10. electrolytic machining 电解加工
11. electrolytic machining tool 电解加工机床
12. electrolytic universal tool and cutter grinder 电解万能工具磨床
13. electrolytic heat treatment 电解液热处理
14. electrohydraulic forming 电液成形
15. electrolytic deburring 电解去毛刺
16. electrolytic marking machine 电解刻印机
17. electrolytic surface grinder 电解平面磨
18. electrochemical corrosion 电化学腐蚀
19. chemical milling 化学铣
20. electrochemical grinding 电化学磨削
21. chemical-mechanical polishing (CMP) 机械化学抛光
22. chemical vapor deposition (CVD) 化学气相沉积
23. electrochemical discharge machining (ECDM) 电化学火花加工
24. electroabrasion 电腐蚀加工法
25. electroplishing 电解抛光

Reading Materials

Chemical Etching Machines

In contrast to the electrochemical processes, the chemical etching process does not use a forming tool nor an external electric power. The work material is mainly removed as a result of differences of potential barriers at the grain boundaries according to the particular material being worked. A variety of acids are used as activators. Filters or centrifuges separate the removed material from the etching medium.

The work is carried out either in a bath of the etching medium (dip etching) or by spraying the etching medium onto the work (spray etching) . In order to obtain a particular work geometry, the parts of the surface which are not to be machined are protected by masking. Frequently, a photographic film technique is employed, whereby the workpiece is covered with a light-sensitive film by rolling it on or dipping. A photographic image of the desired work geometry is then projected onto the work surface, so that the illuminated areas of the work become sensitive to the acid attack and the remaining areas are suitably masked.

Ultrasonic Machining Installations

Ultrasonic machining installations are mainly used for machining electrically non-conductive, brittle materials (such as glass, ceramic oxides, precious stones, carbides, germanium, silicons, graphite and hard metals) . A high frequency generator activates the magnetostrictive oscillator, which transmits the high-frequency oscillations to the tool soldered to the tapered bronze transformer. The tool itself is only indirectly active in the actual metal removal process. The work material is removed through abrasive grains suspended in a slurry, which acts in a manner similar to that of the lapping process—like a number of simultaneously acting chisel points. The slurry suspension is externally applied to the work area and sucked up through the transformer.

In many installations, no separate feeding mechanism is provided. By a vertical arrangement of the tool-work system, the tool advances into the work as a result of its own weight.

Unit 3 Electrical Discharge Machines

Text

When applying the electrical *discharge* machining (EDM) process, the material is eroded as a result of an electrical discharge between tool and workpiece. Due to the *resultant* short-lived, but very high, temperature rises, metal particles at the point of discharge are molten, partially *vaporized* and removed from the melt by mechanical and electromagnetic forces. The working medium is a *dielectric*, which washes the eroded material away and simultaneously acts as a coolant.

As in the case of ECM, EDM is a copying process where there is no contact between tool and workpiece. *Contrary* to ECM, however, there is some erosion of the tool in EDM, which must be allowed for in the tool design to ensure accuracy of machining.

A further difference arises from the fact that in EDM there is no fixed tool feed, but the gap size must be maintained in accordance with the rate of metal removal and the conditions existing within the gap①.

Electrodischarge Die-sinking Machines. The construction principles of an EDM die-sinking machine are shown in Figure 8-3-1. Spark erosion takes place in a container filled with the dielectric, in which the work is clamped. The controlled feed of the electrode is through an electrohydraulic or electromechanical servo system. The electrical energy for erosion is provided by the power generator. The filtering unit separates the eroded material from the dielectric. In the upper left of Figure 8-3-1, a single discharge is illustrated in enlarged form. The applied voltage *ionizes* the gap at the beginning of the discharge. At the point of highest field strength, a channel is formed through which the discharge current flows. At each end of the channel, the material melts and the channel and its *surrounding* gas bubble expand. When the voltage is fully discharged, the channel *collapses* and the molten material vaporizes, simulating a *miniature explosion*. The resultant *crater* is a *hallmark* of the irregular and *scarred* surface finish of spark-eroded work.

Electrodischarge Cutting Machines. An important application of the spark-erosion process is the cutting of metal by wire electrodes. The process is used for the production of *apertures* in cutting tools and the manufacture of tool electrodes for EDM. Figure 8-3-2 illustrates the principle. The cutting tool is a thin copper or brass wire, which enters the work during cutting without physical contact. This *suffers* wear as a result of the action of spark erosion, and for this reason fresh wire is constantly supplied. The *apparatus* required for wire feeding can be seen in Figure 8-3-3. The degree of wire tension, the rate of wire consumption and the reach of the wire support arms are adjusted in accordance with the work to be done and the size of the workpiece②. The working medium for electrodischarge cutting is usually de-ionized water, which is fed to the work area with the use of *flushing* jets.

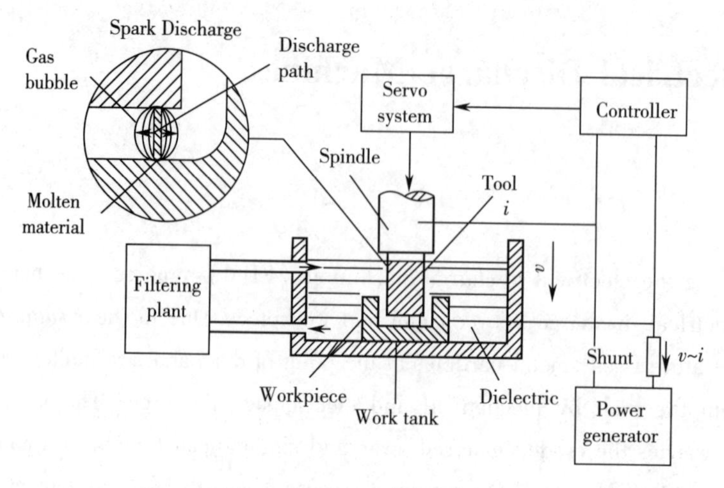

Figure 8-3-1 Electrodischarge erosion plant

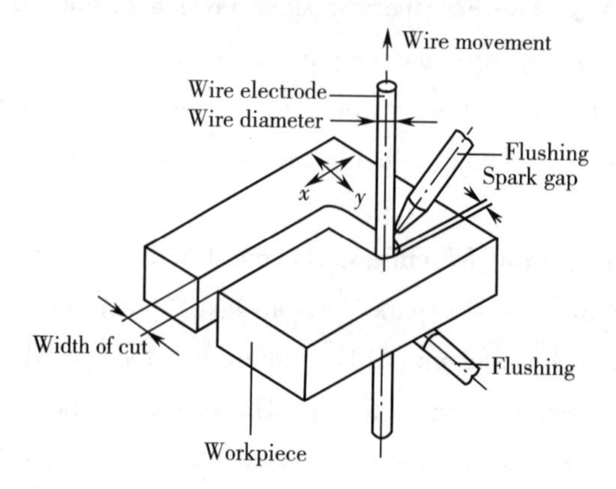

Figure 8-3-2 Numerically controlled EDM cutting x and y right-angled coordinates

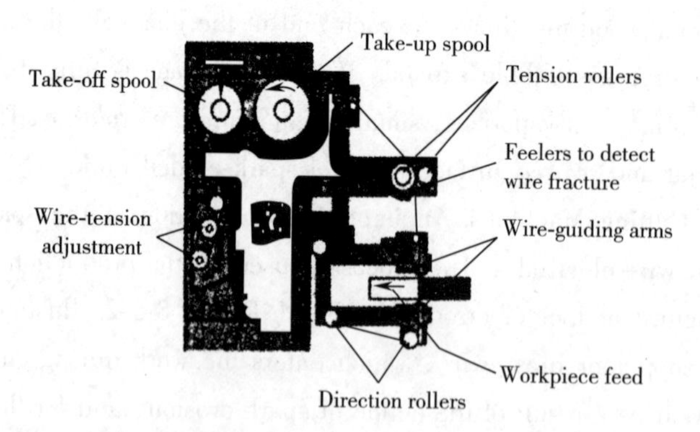

Figure 8-3-3 Wire feed control for an EDM cutting machine

According to the required contour of the workpiece, the table with the workpiece clamped to it and the slide with the wire feed unit must be suitably positioned. The relative advance of the cutting tool to the workpiece does not have a constant velocity, but must be varied in accordance with the conditions existing in the gap throughout the process, depending on the progress of the cut, as was the case in electrodischarge machining.

Questions

1. What are the differences between ECM and EDM?
2. List one of the important applications of the spark-erosion process.
3. Describe the operational principles of an EDM die-sinking machine.

New Words and Expressions

1. discharge [dis'tʃɑːdʒ] v. 排出，释放 n. 放电
2. resultant [ri'zʌltənt] adj. 作为结果而发生的，合成的 n. 结果，组合，合力
3. vaporize ['veipəraiz] v. （使）汽化，（使）蒸发
4. dielectric [daii'lektrik] n. 电介质，绝缘材料 adj. 电介质的，绝缘的
5. contrary ['kɔntrəri] adj. 相反的，逆行的，矛盾的
6. ionize ['aiənaiz] vt. 使电离，离子化
7. surrounding [sə'raundiŋ] adj. 周围的 n. 环境，外界
8. collapse [kə'læps] v. 崩溃，倒塌，破裂
9. miniature ['minjətʃə] n. 缩样，雏形 adj. 小型的，微型的
10. explosion [iks'pləuʒən] n. 爆发，爆炸，迅速增长
11. crater ['kreitə] n. 陷穴，焊口
12. hallmark ['hɔːlmɑːk] n. 检验焰印，品质证明 vi. 在……盖上纯度检验印记
13. scar [skɑː] n. 伤疤，斑疤 vi. 使留下伤痕
14. aperture ['æpətjuə] n. 隙，缝
15. suffer ['sʌfə] vt. 遭受，经受
16. apparatus [ˌæpə'reitəs] n. 机械，设备，装置，仪器
17. flushing ['flʌʃiŋ] n. 冲洗，净化
18. in accordance with 按照，依据，与……一致
19. according to 按照，根据，与……对应

Notes

[1] A further difference arises from the fact that in EDM there is no fixed tool feed, but the gap size must be maintained in accordance with the rate of metal removal and the conditions existing within the gap.

电火花加工与电化学加工之间还有更重要的区别。事实上，电火花加工的刀具进给量不是固定的，加工间隙的大小是依据切削金属的速率与该间隙的加工状态来确定的。

由 that 引出的从句是抽象名词 fact 的同位语从句，它实际上由两个句子组成，中间用“，”隔开；句尾的分词短语 existing within the gap 是 the conditions 的后置定语，它们合起来可译为“加工间隙的工作状态”。

[2] The degree of wire tension, the rate of wire consumption and the reach of the wire support arms are adjusted in accordance with the work to be done and the size of the workpiece.

金属切割丝的张紧程度、丝线损耗速率以及支撑臂的长短是依据所要完成的加工任务及工件尺寸来调整的。

句中三个并列主语 the degree of... , the rate of... , the reach of... 属于“the + 具有动作意义的名词 + of + 名词”的结构，最后一个名词多数是行为对象。

Glossary of Terms

1. spark-erosion machining 电火花加工法
2. electrical discharge machining (EDM) 电火花加工
3. servo system 伺服系统
4. electrodischarge cutting machine 电火花切割机
5. electrical discharge machine 电火花加工机床
6. electrical spark-erosion perforation 电火花打孔
7. electrode contact surface 电极接触面
8. electrical discharge forming 电火花成形机
9. laser cutting machine 激光切割机
10. electron beam cutting machine 电子束切割机
11. cavity sinking EDM machine 型腔电火花加工机床
12. EDM grinders 电火花加工磨床
13. traveling-wire EDM machine 线电极电火花加工机床
14. vertical spark-erosion forming machine 立式电火花形腔加工机床
15. horizontal spark-erosion forming machine 卧式电火花形腔加工机床
16. electron beam machine tool 电子束加工机床
17. tiny hole spark-erosion grinding machine 电火花小孔磨床
18. spark-erosion cutting with a wire 电火花线切割
19. wire cut electric discharge machine 电火花线切割机
20. unidirectional wire travelling 单向走丝运动
21. reciprocating wire travelling 往复走丝运动
22. multi-electrode machining 多电极加工
23. discharge gap 放电间隙
24. electrodischarge wire cutting (EDWC) 电火花线切割加工

Reading Materials

Electron Beam Cutting Installations

In electron beam cutting, the workpiece material is vaporized as a result of a beam of accelerated electrons impinging on a point of contact. Within a few milliseconds, a channel is cut into the workpiece material. The vapour pressure forces the molten metal in the immediate vicinity out of the channel. The depth diameter and form of the cut can be controlled through the characteristics of the beam.

Apart from the actual work chamber, a high-voltage source (up to 150kV) is required, as are devices for the positioning of the electron beam and the workpiece in relation to it. The process takes place in a vacuum in order to avoid the energy-absorbing collision of the electrons with air molecules. The beam may be deflected sideways and focused through the use of a system of magnetic lenses and deflection coils. The power density may be up to 106 kW/cm^2 with a minimum beam diameter of 2μm.

Laser Cutting Machines

Laser cutting uses the erosion effect of high-energy light beams. As in the case of electron beam cutting, the workpiece material is vaporized at the point of impact. According to the laser material used, differentiation is made between solid and gas lasers. With solid lasers (e. g. ruby, neodymium-yttrium-aluminum-garnet), the excitement of a light emission is achieved with the use of a flash light (pump light), and when using a gas laser (e. g. CO_2, He - Ne) through the provision of a high voltage. A lens system focuses the monochromatic high-energy light. The power density achievable at the point of contact with the work may be up to 107kW/cm^2.

What Is Traveling-Wire EDM?

As with all electrical discharge machining, the actual metal removal is the result of a spark discharge jumping the gap from a tool through a dielectric to the workpiece. Upon striking the workpiece, the spark generates an intense localized heat that vaporizes a microscopic portion of the workpiece surface. In time the workpiece is simply eroded away. The traveling-wire method employs a reel of copper wire that is slowly fed past the workpiece and functions much the same as a band saw except the wire is used only once. In jumping from the wire to the workpiece, the sparks erode away a clearly defined path, which can be very closely controlled with NC. Hardened tool steels and carbides are effectively machined with this method. There is only one requirement—the workpiece must be electrically conductive. Traveling-wire EDM is shown in Figure 8-3-4.

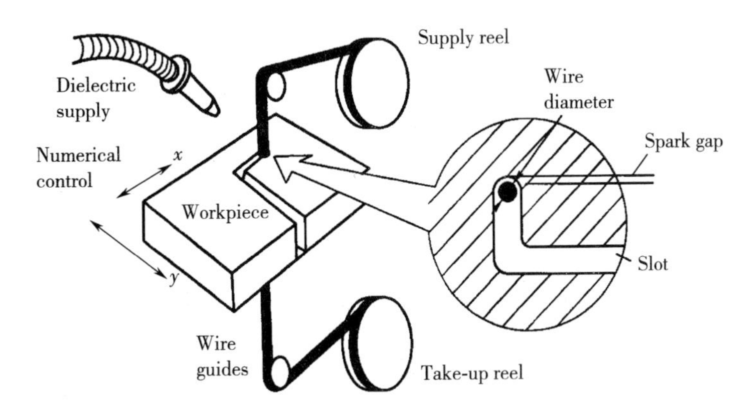

Figure 8-3-4 Traveling-wire electrical discharge machining

Advantages. EDM has many advantages that stem from three basic facts:

The hardness of the workpiece is not a factor. As long as the material can conduct current, it can be machined.

Any shape that can be produced in a tool can be reproduced in the workpiece.

The absence of almost all mechanical force makes it possible to machine the most fragile components without distortion.

Disadvantages. Tool wear requires stepped tooling or redressing of tools for deep holes.

EDM leaves a recast layer at the surface of the cut.

EDM is slow then compared to conventional methods. Whenever possible, the cavities are roughed out prior to heat treatment and then finished by EDM after heat treatment.

Chapter 9 Modern Manufacturing Technology

Unit 1 Rapid Prototyping and Manufacturing

Text

Prototyping (*prototype*) is "the *original* thing in relation to any copy, *imitation*, *representation*, later specimen or improved form" (taken from Webster's dictionary). Rapid prototyping is "fabrication of a physical, three-dimensional part of *arbitrary* shape directly from a numerical *description* (typically a CAD *model*) by a quick, highly automated and totally flexible processes" ("Rapid prototyping and Report", October 1992).

In recent years and constantly being updated at the time of this writing, several types of rapid prototyping and manufacturing (RP&M) have *emerged*. The technologies developed include stereolithography (SLA), selective laser sintering (SLS), fused deposition modeling (FDM), laminated object modeling (LOM), and three dimensional printing (3-D printing), and they are all capable of generating physical objects from computer aided design (CAD) databases.

With these technologies, product development times are substantially decreased and flexibility for manufacturing a variety types and sizes of products is improved.

The features of rapid prototyping and manufacturing

RP&M process consists of two steps: step1, a part is first modeled by a 3-D solid *geometric* modeller, and then is *sliced* into a series of parallel 2-D cross-section layers in computer; step2, the *datum* of the 2-D layers is directly used to instruct the machine for producing the part layer by layer from bottom to top. A common important feature of RP&M is that the prototype part is produced by assigning materials rather than removing materials①.

The benefits from applying the technology to improve product development are in the following three aspects:

(1) Design engineering.

Designers use CAD to generate *visual* models of actual complex products, which are prototypes of products, in short time, therefore engineers can evaluate a design very quickly.

A RP&M prototype can be produced quickly without substantial tooling and labor cost, and the product quality can be improved within the limited time frame and with *affordable* cost②.

(2) Manufacturing.

By providing a physical product at design stage we can speed up process planning and tooling design, reduce problems in interpreting the blue prints on the shop floor.

(3) Marketing.

By *demonstrating* the concept, design ideas, as well as the company's ability to produce it as a prototype, we can gain customer's feedback for design modifications in timely manner, and promote product sales.

Stereolithography

Stereolithography (SLA), which is rapid prototyping by laser curing a photocurable liquid, was launched commercially by 3D systems Inc. in 1987.

SLA apparatus creates the prototype by tracing layer cross-sections on the surface of the liquid resin photopolymer pool with a laser beam. As show in Figure 9-1-1, the laser moves as a point source across the surface of the liquid, first curing the bottom slice of the object. This slice moves down on an elevator by 50 microns to 375 microns (0.002 inch to 0.015 inch, 1 inch =25.4mm), depending on desired accuracy. The next layer is then photocured, also fusing to the one below.

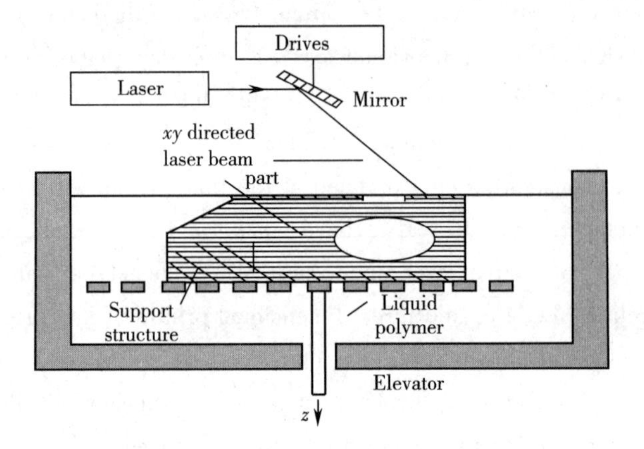

Figure 9-1-1 Stereolithography (SLA), based on commercially published brochures of 3D systems Inc.

The photocurable liquid resin was developed for printing and for furniture *lacquer/sealant*. The laser provides direct energy. With the energy, the original *vinyl monomers* (small molecules) are polymerized into large molecules, which get much more strength from this process.

Unlike the contouring or *zigzag* cutter movement used in CNC machining, in order to create any individual layer, the laser traces out the boundaries of layer first. This is called bordering; imagine a large elastic band or *loop* lying on the surface. Secondly, *hatched* areas are filled in, causing the final *gelling* and solidification. After each layer is formed, the scanning moves to the next layer to draw a new layer on the top of the previous one. In this way, the models built layer by layer from bottom to top. However, some careful process planning is needed to create the accuracy of only a few thousandths of an inch (25.4mm).

When all layers are completed, the prototype is about 95% cured. Post-curing is needed to completely solidify the prototype. This is done in a *fluorescent* oven where *ultraviolet* floods the object (prototype).

Questions

1. What is prototyping?
2. What are the features of rapid prototyping and manufacturing?
3. What is stereolithography (SLA)?
4. Why do we need rapid prototyping and manufacturing?

New Words and Expressions

1. prototype [ˈprəutətaip] n. 原型，模型；样品
2. original [əˈridʒənəl] adj. 最初的，新颖的 n. 原物
3. imitation [imiˈteiʃən] n. 模仿，仿造
4. representation [reprizenˈteiʃən] n. 代表，表示，描写
5. arbitrary [ˈɑːbitrəri] adj. 随意的，专断的
6. description [disˈkripʃən] n. 描写；图说；绘制
7. model [ˈmɔdl] n. 模型，原型，设计图 vt. 仿造，设计
8. emerge [iˈməːdʒ] vi. 浮现，出现，发生
9. geometric [dʒiəˈmetrik] adj. 几何学的
10. slice [slais] n. 切片，分层 vt. 把……切成薄片
11. datum [ˈdeitəm] n. 材料，资料；数据，论据；基准面
12. visual [vizjuəl] adj. 可见的，光学的
13. afford [əˈfɔːd] vt. 供应得起，抽得出（时间）；提供
14. demonstrate [ˈdemənstreit] vt. 论证，说明，示范
15. lacquer [lækə] n. 漆 vt. 上漆
16. sealant [ˈsiːlənt] n. 密封剂
17. vinyl [ˈvainil] n. 乙烯基；~resin 乙烯基树脂
18. monomer [ˈmɔnəmə] n. 单体
19. zigzag [ˈzigzæg] n. Z 字形，锯齿形 adj. 曲折的
20. loop [luːp] n. 圈，环 vt. 使成圈，做环状运动
21. hatch [hætʃ] n. 孔，天窗，图画阴影线 vt. 画阴影线
22. gel [dʒel] n. 凝胶 vi. 胶化
23. fluorescent [fluəˈresənt] adj. 荧光的
24. ultraviolet [ˌʌltrəˈvaiəlit] adj. 紫外（线）的 n. 紫外线辐射
25. in (with) relation to 关于，涉及，与……有关
26. be filled in 填充，填满，把……插进去
27. trace out 描出……轨迹，轨迹为
28. post-curing 辅助固化
29. layer by layer 一层一层
30. from bottom to top 从底到顶
31. prototype workpiece 样件

Notes

[1] A common important feature of RP&M is that the prototype part is produced by assigning materials rather than removing materials.

快速原型制造的一个共有的重要特征，是通过添加材料而不是去除材料来生产原型零件。

句中 that the prototype part is produced by assigning materials rather than removing materials 为表语从句，可译为“通过添加材料而不是去除材料来生产原型零件”，rather than 作“而不是”解。

[2] An RP&M prototype can be produced quickly without substantial tooling and labor cost, and the product quality can be improved within the limited time frame and with affordable cost.

快速原型制造无需支付大量的工具及劳动成本即可快速制造出原型，同时还能在限定的时间范围内以可负担的成本改进产品的质量。

句中 without substantial tooling and labor cost 和 within the limited time frame and with affordable cost 为介词短语，分别在句子中做状语，without 作“没有”解，within 作“在……内”解，而 with 作“用”解。

Glossary of Terms

1. rapid prototyping（RP） 快速原型（成型）
2. rapid prototyping and manufacturing（RP&M） 快速原型（成型）制造
3. three dimensional printing（3-D printing） 三维打印
4. freeform fabrication 自由制造
5. desk-top manufacturing 桌面制造
6. stereolithography（SL） 光固化立体造型
7. stereolithography apparatus（SLA） 光固化成型机
8. selective laser sintering（SLS） 选域（选区）激光烧结
9. fused deposition modeling（FDM） 熔融沉积成型
10. laminated object modeling（LOM） 分层实体制造（叠层制造）
11. laser technology 激光技术
12. dislodge forming 去除成型（车、铣、刨、磨、钻等）
13. additive forming 添加成型（如快速成型）
14. forced forming 受迫成型（如铸造、锻造、粉末冶金等）
15. growth forming 生长成型（利用生物材料的活性，如“克隆”）
16. computed tomography（CT） 断层扫描
17. nuclear magnetic resonance（NMR） 核磁共振
18. high temperature sintering 高温烧结
19. direct metal deposition（DMD） 直接金属沉积
20. rapid tooling 快速模具（工具）制造

Reading Materials

Selective Laser Sintering (SLS)

Selective laser sintering (SLS) was first commercialized by DTM corporation, and is rapid prototyping by laser tracing the shape of the part to be modeled in a thin layer of polymer or ceramic powder, and sintering (softening and bonding) of the powders.

In this process the laser moves as a point source across the surface of the powder, first sintering the bottom slice of the desired object. A roller spreads more power and a second layer is sintered, also fusing to the one below. This process is repeated over layers of powder. Figure 9-1-2 shows the working principle of SLS.

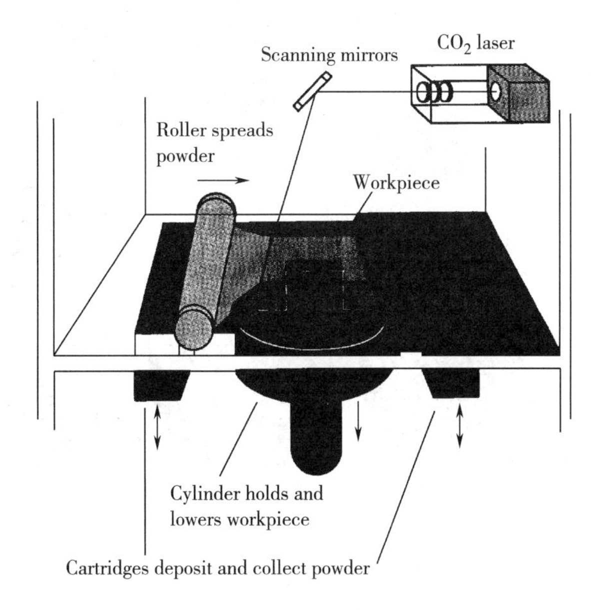

Figure 9-1-2 Selective laser sintering (SLS), based on commercially published brochures from the DTM corporation

In many respects, SLS is similar to SLA except that the laser is used to sinter and fuse powder rather than photocure a polymeric liquid, and a thin layer of fusible powder is heated by infrared heating panels. Otherwise comparing with SLA, this process can rely on the supporting strength of the unfused powder around the partially fused object. Therefore, support columns for any overhanging parts of the component are not needed. This allows the creation of rather delicate, lacelike objects. Nevertheless SLS parts have a rough, grainy appearance from the sintering process, and it is often preferable to hand smooth the surfaces.

Laminated Object Modeling (LOM)

Laminated object modeling (LOM) was developed by Helisys Inc., and was first offered commercially in the period from 1987 to 1990, and is rapid prototyping by laser cutting the top layer of a stack of paper, each layer of which is glued down.

LOM processes produce parts from bonded paper plastic, metal, or composite sheet stock. LOM machines bond a layer of sheet material to a stack of previously formed laminations, and then a laser beam cuts the outline of the part cross-section generated by CAD to the required shape. The layers can be glued or welded together and the excess material of every sheet either is removed by vacuum suction or remains as the next layer's support. This process is repeated by using very thin layers of material. Figure 9-1-3 shows the working principle of LOM.

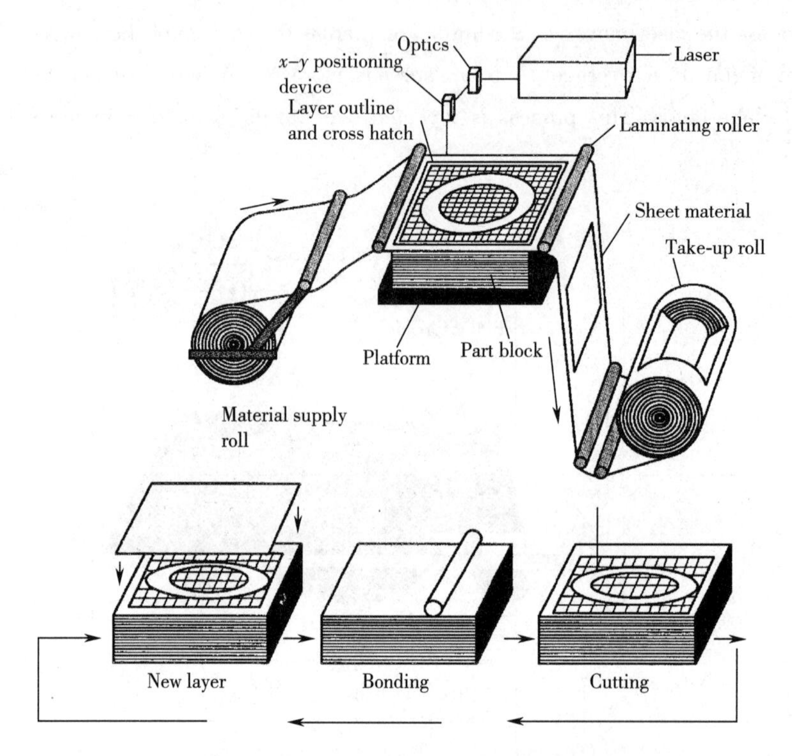

Figure 9-1-3 Laminated object modeling (LOM), based on commercially published brochures of Helisys Inc.

For larger components especially in the automobile industry, LOM is often preferred to the SLA or SLS processes.

Fused Deposition Modeling (FDM)

Fused deposition modeling (FDM), which constructs parts based on deposition of extruded thermoplastic materials, was developed by Stratasys Inc. and is executed on machines called the FDM 1620, 2000, or 8000 series.

The FDM process is that the thermoplastic modeling material is fed into the temperature-controlled FDM extrusion head and heated to a liquid state. The head extrudes and deposits the material in ultrathin layers onto a fixtureless base.

Figure 9-1-4 shows that the material is supplied as a filament from a spool. The overall geometry and system are reminiscent of icing a cake. The ribbon through an exit nozzle is guided by computer, and the viscous ribbon of polymer is gradually built up from a fixtureless base plate. In terms of motion control, FDM is more similar to CNC machining than SLA or SLS. For simple parts, there is no need for fixing, and material can be built up layer by layer. The creation of more complex parts with inner cavities, unusual sculptured surfaces and overhanging features does require a support base, but the supporting material can be broken away by hand, thus requiring minimal finishing work.

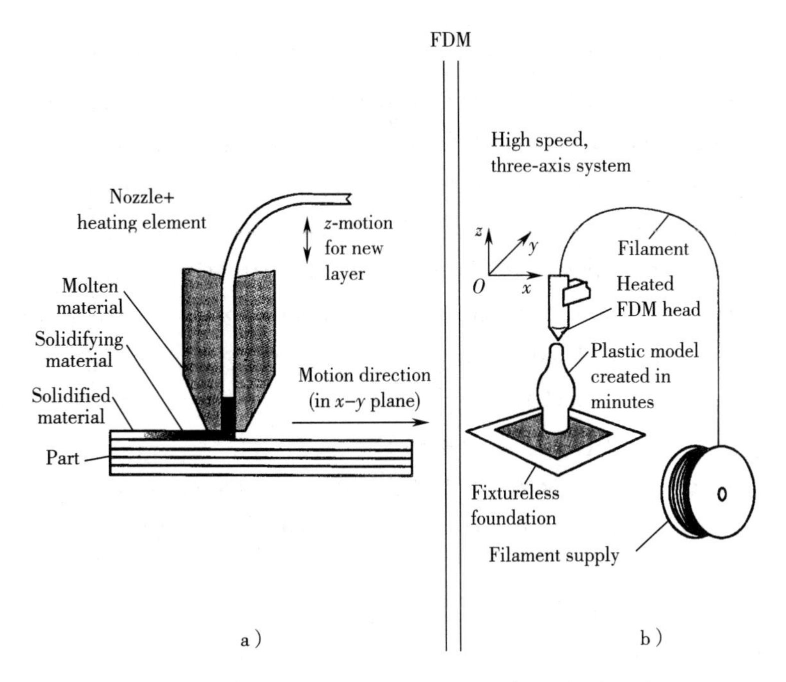

Figure 9-1-4 Fused deposition modeling (FDM), based on published brochures of Stratasys Inc.

Unit 2 Advanced Manufacturing Technology

Text

Flexible Manufacturing. In the modern manufacturing *setting*, *flexibility* is an important characteristic. It means that a manufacturing system is *versatile* and *adaptable*, while also capable of *handling* relatively high production runs. A flexible manufacturing system is versatile in that it can produce a variety of parts. It is adaptable because it can be quickly *modified* to produce a completely different line of parts.

A flexible manufacturing system (FMS) is an *individual* machine or a group of machines served by an automated material handling system that is computer controlled and has a tool handling capability①. Because of its tool handling capability and computer control, such a system can be continually *reconfigured* to manufacture a wide variety of parts. This is why it is called a flexible manufacturing system. A FMS typically *encompasses*:

(1) Process equipment, e. g. machine tools, assembly stations, and *robots*.

(2) Material handling equipment, e. g. robots, *conveyors*, and AGVs (automated guided vehicles).

(3) A communication system.

(4) A computer control system.

Flexible manufacturing represents a major step toward the goal of fully integrated manufacturing. It involves integration of automated production processes. In flexible manufacturing, the automated manufacturing machine (i. e., lathe, mill, drill) and the automated material handling system *share instantaneous* communication *via* a computer network②.

Flexible manufacturing takes a major step toward the goal of fully integrated manufacturing by integrating several automated manufacturing concepts:

(1) Computer numerical control (CNC) of individual machine tools.

(2) *Distributed* numerical control (DNC) of manufacturing system.

(3) Automated material handling systems.

(4) Group technology (families of parts).

When these automated processes, machines and concepts are brought together in one integrated system, a FMS is the result. Humans and computers play major roles in a FMS. The amount of human labor is much less than a manually operated manufacturing system, of course. However, humans still play a *vital* role in the operation of a FMS. Human tasks include the following:

(1) Equipment *troubleshooting*, *maintenance*, and repair.

(2) Tool changing and setup.

(3) Loading and unloading the system.

(4) Data input.

(5) Changing of part programs.

(6) Development of programs.

Flexible manufacturing system components. A FMS has four major components: machine tools; control system; material handling system; human operators.

Questions

1. What is a FMS?
2. What are human tasks in a FMS?
3. What are the four major components of a FMS?
4. Why do we need advanced manufacturing technology?

New Words and Expressions

1. setting [ˈsetiŋ] n. 安装，装置，定位
2. flexibility [fleksəˈbiliti] n. 柔性，适应性，灵活性
3. versatile [ˈvəːsətail] adj. 多用途的，万用的，通用的
4. adaptable [əˈdæptəbl] adj. 适合的，可适应的
5. handling [ˈhændliŋ] n. 掌握，操作，处理
6. modified [mɔdifaid] adj. 改进的；变形的
7. individual [indiˈvidjuəl] adj. 单独的，个别的 n. 个体
8. reconfigure [riːkənˈfigə] v. 重新装配，改装
9. encompass [inˈkʌmpəs] vt. 包围，包含，完成
10. robot [ˈrəubɔt] n. 自动机，机器人
11. conveyor [kənˈveiə] n. 运送者，传送装置，传送带
12. share [ʃεə] n. 共享，参与 v. 分享，共有
13. instantaneous [instənˈteinjəs] adj. 同时的，瞬间的
14. via [ˈvaiə] prep. 经由，经过
15. distribute [disˈtribju（ː）t] vt. 分发，分配，分类 v. 分发
16. vital [ˈvaitl] adj. 生命的，主要的 n. 生机，要害
17. troubleshooting [ˈtrʌblʃuːtiŋ] n. 故障检修，排除故障
18. maintenance [ˈmeintinəns] n. 维持，保持，维修
19. be vital to 是……所必需的，对……极重要的
20. a wide variety of 各种各样的，各种类型的
21. play a vital role in (play major roles in) 在……中起重要（主要）作用
22. a major step toward 迈向……的重要一步
23. production runs 流水线生产
24. communication system 通信系统

Notes

[1] A flexible manufacturing system (FMS) is an individual machine or a group of machines served by an automated material handling system that is computer controlled and has a tool handling capability.

柔性制造系统（FMS）是指一台机器或一组机器，它们的服务系统是由计算机控制，并具有工具操作能力的自动化物料输送系统。

句中 that is computer controlled and has a tool handling capability 为由 that 引导的定语从句，修饰 an automated material handling system。翻译时先翻译从句，后翻译主句。

[2] In flexible manufacturing, the automated manufacturing machine (i. e., lathe, mill, drill) and the automated material handling system share instantaneous communication via a computer network.

在柔性制造过程中，自动化的制造机器（即车床，铣床，钻床）和自动化的物料输送系统同时共享经由计算机网络传送的信息。

句中 in flexible manufacturing 为介词短语做状语，the automated manufacturing machine (i. e., lathe, mill, drill) and the automated material handling system 为本句的主语。

Glossary of Terms

1. man machine engineering (MME) 人机工程
2. group technology (GT) 成组技术
3. advanced manufacturing technology 先进制造技术
4. flexible manufacturing cell (FMC) 柔性制造单元
5. flexible manufacturing system (FMS) 柔性制造系统
6. automated guided vehicle system (AGVS) 自动导向小车系统
7. computer-integrated manufacturing (CIM) 计算机集成制造
8. agile manufacturing (AM) 敏捷制造（灵捷制造）
9. virtual manufacturing (VM) 虚拟制造
10. environmentally conscious design and manufacturing (ECDM) 环保设计与制造
11. environmental engineering 环境工程
12. nanomaterial 纳米材料
13. nanotechnology 纳米技术
14. reverse engineering (RE) 反求工程
15. computer-integrated production management (CIPM) 计算机集成生产管理
16. concurrent engineering (CE) 并行工程
17. just-in-time (JIT) 即时生产（精益生产，准时生产）
18. clean production (CP) 清洁化生产
19. lifecycle engineering (LCE) 生命周期工程
20. design for environment (DFE) 面向环境的设计

21. green product design (GPD) 绿色产品设计
22. intelligent manufacturing system (IMS) 智能制造系统
23. multi-media technology (MMT) 多媒体技术
24. chemical vapor deposition (CVD) 化学气相沉积
25. automatic storage and retrieval system (AS/RS) 自动化仓库系统
26. bionic manufacturing (BM) 仿生制造
27. industrial robot 工业机器人

Reading Materials

Computer Integrated Manufacturing (CIM)

Computer integrated manufacturing (CIM) is the term used to describe the modern approach to manufacturing. Although CIM encompasses many of the other advanced manufacturing technologies such as computer numerical control (CNC), computer-aided design/computer-aided manufacturing (CAD/CAM), robotics, and just-in-time delivery (JIT), it is more than a new technology or a new concept. CIM is an entirely new approach to manufacturing, a new way of doing business.

To understand CIM, it is necessary to begin with a comparison of modern and traditional manufacturing. Modern manufacturing encompasses all of the activities and processes necessary to convert raw materials into finished products, deliver them to the market, and support them in the field. These activities include the following:

(1) Identifying a need for a product.

(2) Designing a product to meet the needs.

(3) Obtaining the raw materials needed to produce the product.

(4) Applying appropriate processes to transform the raw materials into finished products.

(5) Transporting products to the market.

(6) Maintaining the product to ensure proper performance in the field.

Fully integrated manufacturing firms realize a number of benefits from CIM:

(1) Product quality increases.

(2) Lead times are reduced.

(3) Direct labor costs are reduced.

(4) Product development times are reduced.

(5) Inventories are reduced.

(6) Overall productivity increases.

(7) Design quality increases.

Group Technology (GT)

Group technology is a concept at the very heart of CIM. CIM is supposed to give manufacturers the flexibility to produce customized products without sacrificing productivity.

Group technology is a key ingredient in the larger formula of CIM that makes this possible. It amounts to making batch manufacturing economical.

GT involves grouping parts according to design characteristics, the processes used to produce them, or a combination of these. Figure 9-2-1 is an example of similar parts grouped together on the basis of two design characteristics (shape and material). These similar parts fall into a family of parts. Families of parts can be produced using similar processes usually within a single flexible manufacturing cell (FMC).

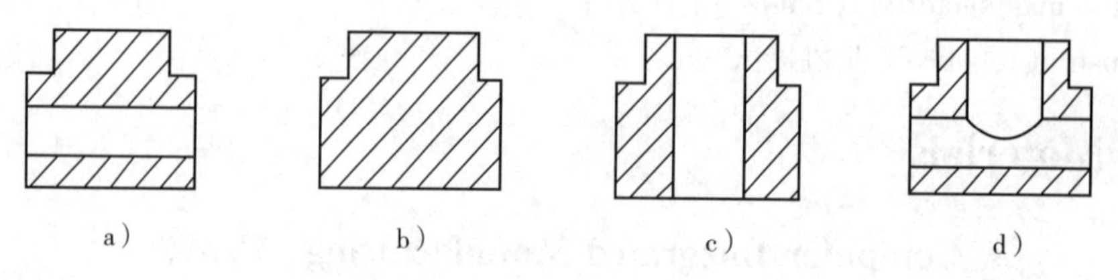

Figure 9-2-1 Parts with similar design

Agile Manufacturing (AM)

Agile manufacturing is a means of thriving in an environment of continuous change, by managing complex inter-and intra-firm relationships through innovations in technology, information, communication, organizational redesign and new marketing strategies.

There are four principles of agility:

(1) To organize to master change ("an agile company is organized in a way that allows it to thrive on change and uncertainty").

(2) To leverage the impact of people and information in an agile company, knowledge is valued, innovation is rewarded, and authority is distributed to the appropriate level of the organization. There is a climate of mutual responsibility for joint success.

(3) To cooperate to enhance competitiveness. Cooperation internally and with other companies is an agile competitor's operational strategy of first choice.

(4) To enrich the customer. An agile company is perceived by its customers as enriching them in a significant way, not only itself.

From everything mentioned above, we can conclude that AM involves more than just manufacturing. It involves the firm's organizational structure, the way in which the firm treats its people, the partnerships with other organizations, and the relationship with customers.

Environmentally Conscious Design and Manufacturing (ECDM)

Environmentally conscious design and manufacturing (ECDM) is a view of manufacturing that includes the social and technological aspects of the design, synthesis, processing, and use of products in continuous or discrete manufacturing industries. The benefits of ECDM include safer and cleaner factories, worker protection, reduced future costs for disposal, reduced environmental and health risks, improved product quality at lower cost, better public image, and higher productivity. Environmentally conscious technologies and design practices will also allow manufacturers to minimize waste and to turn waste into a profitable product.

Research on ECDM can be categorized into two areas, namely, environmentally conscious product design and environmentally conscious process design, also called environmentally conscious manufacturing (ECM).

The research issues in ECDM include: integration of product and process design with material selection systems; development of models for assessing the integration of consumer demand and product use, disposal or recycling; improvement in methods, tools and procedures for evaluation of the risks associated with environmental hazards and the cost or benefit; substitution of materials with lower environmental impact in processing or in the final product; advancement in techniques for forecasting the effects of specific governmental regulations over the complete product lifecycle; new or improved manufacturing processes, and development of new bulk materials and coatings with increased life spans that can be manufactured with decreased environmental impact.

ECDM is a long-term goal that will require changes in management systems and cost models (economic, financial and accounting). To achieve this goal, controlling and monitoring of operations must be increased to better comprehend environmental costs.

Nanomaterial and Nanotechnology

Nanomaterials and nanotechnology have become a magic word in modern society. Nanomaterials represent today's cutting edge in the development of novel advanced materials which promise tailor-made functionality and unheard applications in all key technologies. So Nanomaterials are considered as a great potential in the 21th century because of their special properties in many fields such as optics, electronics, magnetics, mechanics and chemistry. These unique properties are attractive for various high performance applications. Examples include wear resistant surfaces, low temperature sinterable high-strength ceramics, and magnetic nanocomposites. Nanostructured materials present great promises and opportunities for a new generation of materials with improved and marvelous properties.

What is nanotechnology? It is a term that entered into the general vocabulary only in the late 1970's, mainly to describe the metrology associated with the development of X-ray, optical and other very precise components. We defined nanotechnology as the technology where dimensions and tolerances in the range 0.1 ~ 100nm (from the size of the atom to the wavelength of light) play a critical role.

Unit 3 High-Speed Machining

Text

With continuing demands for higher productivity and lower production costs, *investigations* have been carried out since the late 1950s to increase the cutting speed and the material-removal rate in machining, particularly for applications in the aerospace and automotive industries.

The term "high" in high-speed machining (HSM) is somewhat relative; as a general guide, an approximate range of cutting speeds may be defined as follows:

High speed: 600 ~ 1800m/min

Very high speed: 1800 ~ 18000m/min

Ultrahigh speed: >18000m/min

Spindle rotational speeds in machine tools now range up to 50000rpm, although the automotive industry generally has limited them to 15000rpm for better reliability and less *downtime* should a failure occur①. The spindle power required in high-speed machining is generally on the order of 0.004 W/rpm—much less than in traditional machining, which is typically in the range of 0.2 ~ 0.4W/rpm. Feed rates in high-speed machining are now up to 1m/s and the *acceleration* rates of machine-tool components are very high.

Spindle designs for high speeds require high stiffness and accuracy and generally involve an *integral* electric motor. The *armature* is built onto the shaft, and the stator is placed in the wall of the spindle housing. The bearings may be rolling elements or *hydrostatic* the latter is more desirable because it requires less space than the former. Because of *inertia* effects during the acceleration and deceleration of machine-tool components, the use of lightweight materials (including ceramics and composite materials) is an important consideration.

The selection of appropriate cutting-tool materials is, of course, a major consideration. It is apparent that (depending on the workpiece material) multiphase coated carbides, ceramics, cubic-boron nitride and diamond are all *candidate* tool materials for this operation.

These studies have indicated that high-speed machining is economical for certain specific applications. For example, it has been implemented in machining ①aluminum structural components for aircraft; ②submarine propellers of 6m in diameter, made of a nickel-aluminum-bronze alloy, and weighing 55000kg; and ③automotive engines, with five to ten times the productivity of traditional machining. High-speed machining of complex three- and five-axis contours has been made possible by advances in CNC technology.

Another major factor in the adoption of high-speed machining has been the requirement to further improve dimensional tolerances in cutting operations. As the cutting speed increases, more and more of the heat generated is removed by the chips, thus the tool and (more importantly) the workpiece remain close to *ambient* temperature[②]. This is beneficial, because there is no thermal expansion or warping of the workpiece during machining.

The important machine-tool characteristics in high-speed machining may be summarized as follow:

(1) Spindle design for stiffness, accuracy and balance at very high rotational speeds.

(2) Bearing characteristics.

(3) Inertia of the machine-tool components.

(4) Fast feed drives.

(5) Selection of appropriate cutting tools.

(6) Processing parameters and their computer control.

(7) Workholding devices that can withstand high centrifugal forces.

(8) Chip removal systems effective at very high rates of material removal.

Questions

1. What is high-speed machining?
2. What are the important machine-tool characteristics in high-speed machining?
3. What is high-speed dry machining?

New Words and Expressions

1. investigation [inˌvestiˈgeiʃən] n. 研究，调查
2. downtime [ˈdauntaim] n. （工厂、机器等）停工期，停机时间
3. acceleration [ækˌseləˈreiʃən] n. 加速，促进，加速度
4. integral [ˈintigrəl] adj. 组成的，完全的 n. 总体，整体
5. armature [ˈɑːmətjuə] n. 转子，电枢，电容板
6. hydrostatic [ˌhaidrəuˈstætik] adj. 静力学的，流体静力学的
7. inertia [iˈnəːʃjə] n. 惯性，惯量
8. candidate [ˈkændidit] n. 选择物，候选人
9. ambient [ˈæmbiənt] adj. 周围的，围绕着的 n. 周围环境
10. on the order of 属于……同类的，约为
11. in the range of 在……范围内，在射程内
12. of course 当然，自然
13. make an investigation on (of , into) sth. 对某事进行调查研究

Notes

[1] Spindle rotational speeds in machine tools now range up to 50000 rpm, although the automotive industry generally has limited them to 15000rpm for better reliability and less downtime should a failure occur.

尽管为了可靠性更好和突发故障时的停机时间更短，汽车工业通常将机床主轴转速限制

在15000r/min之内，但目前机床主轴的转速可以达到50000r/min。

句中for为表示目的的介词，可译为“为了”; should a failure occur意为“万一发生故障”。

[2] As the cutting speed increases, more and more of the heat generated is removed by the chips, thus the tool and (more importantly) the workpiece remain close to ambient temperature.

当切削速度增加时，所产生的大部分热量由切屑带走，因此刀具和（更重要的是）工件可保持接近于周围环境的温度。

句中as the cutting speed increases为as引导的时间状语从句，意为“正当……的时候；随着……”; close to可译为“接近于；在附近”。

Glossary of Terms

1. high-speed machining 高速切削
2. high-speed dry machining 高速干切削
3. hard machining 硬切削
4. ultrahigh-speed machining 超高速切削
5. ultraprecision machining 超精切削
6. flush-fine machining 流体精加工
7. material removal rate 材料切除率
8. high-speed cutting technology 高速切削技术
9. high-speed lathe 高速车床
10. high-speed grinding 高速磨削
11. high speed tool steel 高速工具钢
12. high-speed cutting machine 高速切削机床
13. high-speed cutting tool 高速切削刀具
14. precision machining 精密切削
15. virtual machining 虚拟切削
16. non-chip finish 无屑加工
17. one-pass machining “一次过”技术
18. high-speed end milling 高速立（端）铣
19. high-speed CNC machining center 高速数控加工中心
20. complex product 复杂产品
21. high-speed grinding wheel 高速砂轮
22. high-speed lubrication 高速润滑
23. high-speed drilling machine 高速钻床

Reading Materials

Hard Machining

We have noted that as the hardness of the workpiece increases, its machinability decreases, and tool wear and fracture, surface finish, and surface integrity can become significant problems. There are several other mechanical (including abrasive machining), nonmechanical and advanced techniques of removing material economically from hard or hardened metals and alloys. However, it is still possible to apply traditional machining processes to hard metals and alloys by selecting an appropriate hard-tool material and using machine tools with high stiffness, power and precision.

A common example is the finish machining of heat-treated steel (45 HRC to 65 HRC) shafts, gears, pinions, and various automotive components using polycrystalline cubic-boron nitride (PCBN), cermet, or ceramic cutting tools. Called hard machining or hard turning, this process produces machined parts with good dimensional accuracy, surface finish (as low as 0.25 μm), and surface integrity. The available power, static and dynamic stiffness of the machine tool and its spindle, and workholding devices and fixtures are important factors.

From technical, economic and ecological considerations, hard turning can compete successfully with the grinding process and with the quality and characteristics approaching those produced in grinding machines.

High-speed Dry Machining

High-speed machining has clear economic advantages as long as spindle damage and chatter are avoided. Similarly, dry machining is recognized as a promising approach to solving environmental problems associated with cutting fluids. Unfortunately, it often has been perceived that the two approaches cannot be applied simultaneously because the higher speeds would result in low tool life. However, high-speed dry machining is possible with aluminum and cast iron and is examined here for roughing, finishing and holemaking operations on a cast iron cylinder head for a 4.3 liter engine.

High-speed dry machining is a demanding operation. Heat dissipation without a liquid coolant requires high-performance coatings and heat-resistant tool materials. The cutting-tool materials are either ceramic or CBN because of the intense heat generated in the process. In addition, machine tools with high power are required, and spindles with high-speed machining, hence CNC programs must be developed that maintain process parameters within a window of operation that is known to be chatter-free.

The key to dry machining the cast-iron head is the use of two advanced techniques. The first is "Red-crescent" machining, where high temperatures are generated in the workpiece in front of the tool, creating a visible red arc (hence the name of the process). The second technique is flush-fine machining, using in spindle high-pressure air to remove chips which normally would be carried away by cutting fluid.

Chapter 10 Qualities of Machined Surface

Unit 1 Measurement and Inspection

Text

The basic purpose of manufacturing is to produce engineering materials and products with specified shapes, sizes and finishes. These shapes, sizes and finish specifications are generally found on the part drawing or the manufacturing drawing, and they are often referred to as quality characteristics①.

Interchangeable manufacture. Our modern mass-production systems, based on the concepts of *interchangeable* manufacture, require that each part or assembly going into a final product be made to definite size, shape and finish specifications. The mass production of both consumer and producer goods relies on *interchangeability*, and interchangeability requires *fabrication* to exacting dimensions and close tolerances. Compressor pistons, for example, must be machined within limits so that a piston selected at random will fit and function properly in the compressor model for which it was designed②. This interchangeable manufacturing system is to a large degree responsible for the high standard of living enjoyed today by the people of the United States. It makes possible the standardization of products and methods of manufacturing and provides for ease of assembly and repair of products.

Quality *assurance*. The quality of a product may be stated in terms of a measure of the degree to which it *conforms* to specifications and standards of workmanship. These specifications and standards should reflect the degree to which the product satisfies the wants of a particular customer or user. The quality assurance function is charged with the responsibility of maintaining product quality consistent with those requirements and it involves the following four basic activities:

(1) Quality specification.

(2) Inspection.

(3) Quality analysis.

(4) Quality control.

Quality specification. The product design engineer provides the basic specifications of product quality by means of the various dimensions, tolerances, and other requirements *cited* on the engineering drawings. These basic specifications are further refined and *elaborated* on the manufacturing and fabrication procedures. In many cases, the quality assurance group will provide a further *interpretation* of those specifications as a basis for specifying inspection procedures.

Inspection. The mass production of interchangeable products is not altogether effective without some means of *appraising* and controlling product quality. Production operators are required, for example, to machine a given number of parts to the specifications shown on a drawing. There are often many factors that cause the parts to *deviate* from specifications during the various manufacturing operations. A few of these factors are variations in raw materials, *deficiencies* in machines and tools, poor methods, excessive production rates, and human errors.

Provisions must be made to *defect* errors so that the production of faulty parts can be stopped. The inspection department is usually charged with this responsibility, and its job is to interpret the specifications properly, inspect for conformance to those specifications, and then convey the information obtained to the production people, who can make any necessary corrections to the process③. Inspection operations are performed on raw materials and purchased parts at "receiving inspection". "In-process inspection" is performed on products during the various stages of their manufacture, and finished products may be subjected to "final inspection".

Quality analysis. During the various inspection processes, a variety of quality information is recorded for review by representatives from both manufacturing and quality assurance. In essence, this information provides a basis for analyzing the quality of the product as it relates to the capability of the manufacturing process and to the quality specifications initially prescribed for the product. Generally, it is not possible to mass produce products that are 100% free of defects except at considerable expense. The quality analysis will determine the level of defectivity of the product so that a decision can be made whether or not that level is tolerable. If a given level of defectivity is not tolerable, decisions must be made relative to what to do with the process, and what to do with the product.

Quality control. The word "control" implies regulation, and of course, regulation implies observation and manipulation. Thus a pilot flying an aircraft from one city to another must first set it on the proper course heading and then must continue to observe the progress of the craft and manipulate the controls so as to maintain its flight path in the proper direction. Quality control in manufacture is an *analogous* situation. It is simply a means by which management can be assured that the quality of product manufactured is consistent with the quality-economy standards that have been established. The word "quality" does not necessarily mean "the best" when applied to manufactured products. It should imply "the best for the money".

Questions

1. What is meant by the term "interchangeable manufacture"?

2. Name and describe the four basic activities commonly involved in the quality assurance function.

3. Why is the inspection process necessary in most mass-production activities?

4. During what stages in the manufacturing process are inspection operations commonly performed?

5. What is meant by the term "quality analysis"?

6. What is meant by the term "quality control"?

New Words and Expressions

1. interchangeable [intəˈtʃeindʒəbl] adj. 可交换的，可互换的
2. interchangeability [intə (ː) ˌtʃeindʒəˈbiliti] n. 互换性
3. fabrication [ˌfæbriˈkeiʃən] n. 制配，生产，建造，构造
4. assurance [əˈʃuərəns] n. 确信，把握，保证
5. conform [kənˈfɔːm] v. 使一致，符合
6. cite [sait] vt. 引用，引证
7. elaborate [iˈlæbərit] adj. 精心制成的，精巧的 vt. 钻研，推敲
8. interpretation [intəːpriˈteiʃən] n. 解释，说明，分析
9. appraise [əˈpreiz] vt. 评价，估价，鉴定
10. deviate [ˈdiːvieit] vi. 超越，偏离 vt. 使背离
11. deficiency [diˈfiʃənsi] n. 不足，缺乏，缺陷
12. defect [diˈfekt] n. 缺点，缺陷
13. analogous [əˈnæləgəs] adj. 类似的，相似的，模拟的
14. be referred to as 称为……；被认为是……
15. in essence 本质上，大体上
16. conform to 使符合，使一致
17. give an assurance that 保证……
18. cite a case (an instance) 举个例子
19. be analogous to 与……相似

Notes

[1] These shapes, sizes and finish specifications are generally found on the part drawing or the manufacturing drawing, and they are often referred to as quality characteristics.

一般（通常）是将零件的形状、尺寸和精度要求标注在零件图或制造图上，并且通常把它们称为零件的质量特征。

句中 be generally found on 是被动语态，可译为“一般（通常）在……上能找到”“一般

地（通常地）被标注在……上”；be often referred to as 可译为“通常称为……”“通常被认为是……”。

［2］Compressor pistons, for example, must be machined within limits so that a piston selected at random will fit and function properly in the compressor model for which it was designed.

如压缩机活塞必须在限定的（尺寸）范围内加工，以便随机选用的活塞能够在相应的压缩机模型中正确安装及运行。

句中 so that a piston selected at random will fit and function properly 为目的状语从句，可译为“以便随机选用的活塞能够正确安装及运行”；for which it was designed 为“介词 + 关系代词”引导的定语从句，修饰 the compressor model。

［3］The inspection department is usually charged with this responsibility, and its job is to interpret the specifications properly, inspect for conformance to those specifications, and then convey the information obtained to the production people, who can make any necessary corrections to the process.

检验部门通常负有这项职责，其工作是正确说明（解释）技术要求并检测这些技术要求（说明书、规范、技术条件、清单）的一致性，然后向生产人员通知（传达）所得到的这些信息，从而使生产人员对生产工艺进行必要的修正。

句中 is charged with 可译为“负有，装有”。

Glossary of Terms

1. interchangeable manufacture　可互换性制造
2. quality assurance　质量保证
3. mass production　成批生产，大批生产
4. schematic drawing　示意图
5. failure analysis　故障分析
6. quality analysis　质量分析
7. quality control　质量控制
8. quality standard　质量标准
9. quality index　质量指标
10. quality specification　质量说明书
11. statistical quality control method　统计质量控制方法
12. population distribution　总体分布
13. sample distribution　样品分布
14. normal distribution　正态分布
15. normal curve　正交曲线
16. process control chart　过程（工艺）控制图

17. raw material 原材料
18. finished product 成品
19. rate of finished product 成品率
20. finish size 成品尺寸
21. semi-finished product 半成品
22. inspection report 检验报告
23. inspection on the spot 现场检验
24. inspection manual 检验手册
25. inspection by sampling 取样检验，抽查
26. inspection between processes 工序检验
27. inspection sheet（ticket）检验单
28. manufacturing cost 制造成本
29. statistical analysis 统计分析
30. statistical error 统计误差
31. statistical estimation 统计估计
32. QE, Quality Engineering 品质工程（部）
33. QA, Quality Assurance 品质保证（处）
34. QC, Quality Control 品质管制（课）
35. PD, Product Department 生产部

Reading Materials

Statistical Quality Control

Increasing demands for higher quality of product often result in increased cost of inspection and surveillance of processes and product. The old philosophy of tempting inspect quality into a product is time consuming and costly. It is, of course, more sensible to control the process and make the product correctly rather than rely on having to sort out defective product from good product. Certain statistical techniques have been developed which provide economical means of maintaining continual analysis and control of processes and product. These are known as statistical quality control methods, and many of them are formulated on the following basic statistical concepts:

(1) A population or universe is the complete collections of objects or measurements of the type in which we are interested at a particular time. The population may be finite or infinite.

(2) A sample is a finite group or set of objects taken from a population.

(3) The average is a point or value about which a population or a sample set of measurements tends to cluster. It is a measure of the ordinariness or central tendency of a group of measurements.

(4) Variation is the tendency for the measurements or observations in a population or a sample to scatter or disperse themselves about the average value.

In many cases the sizes observed from the inspection of a dimension of a group of pieces have been found to be distributed as shown in Figure 10-1-1. If the pins represented in Figure 10-1-1 include all the existing pins of a particular type, then this would be referred to as a population distribution. If those pins represent only a portion of a larger batch of the same type pins, then it would be called a sample distribution.

The pattern of variation shown in Figure 10-1-1 is typical of that obtained from data taken from many natural and artificial processes. It is referred to as a normal distribution, and the smooth curve formed by this distribution is called the normal curve.

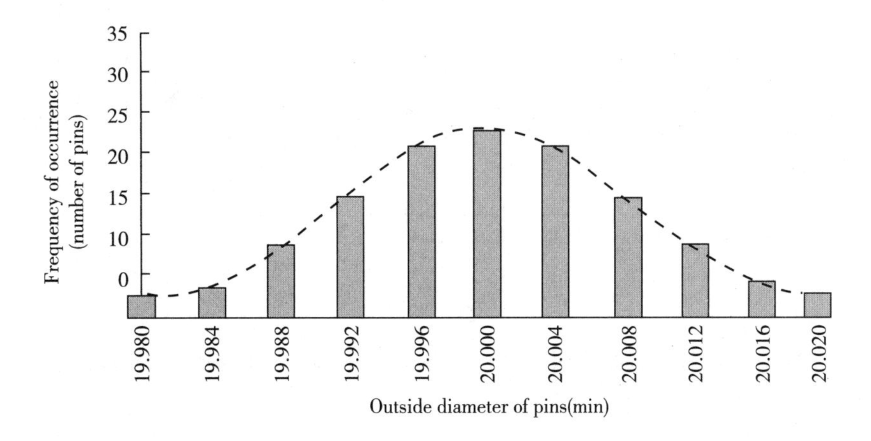

Figure 10-1-1 Frequency distribution of outside diameters of ground pins

Process Control Charts

One of the statistical tools that is commonly used in quality control work to evaluate the process is the control chart. A control chart is simply a frequency distribution with the observed values plotted as points joined by lines in the order of occurrence so that each value retains identity relative to time. The chart is provided with limit lines, called control limits, within which the points fall if influenced only by chance causes.

Two of the most common charts used in process control are the chart for averages, $\overline{X}$-chart, and the chart for ranges, R-chart. Figure 10-1-2 shows an example of such charts for a rough turning operation on the outside diameter of steel shafts. The first point on the $\overline{X}$-chart of Figure 10-1-2 was determined by measuring the diameters of five shafts and then calculating the average of those measured values. Each succeeding point is also an average of the measured diameters of five shafts. Average values are plotted instead of individual values because sample averages tend to be more normally distributed than single values.

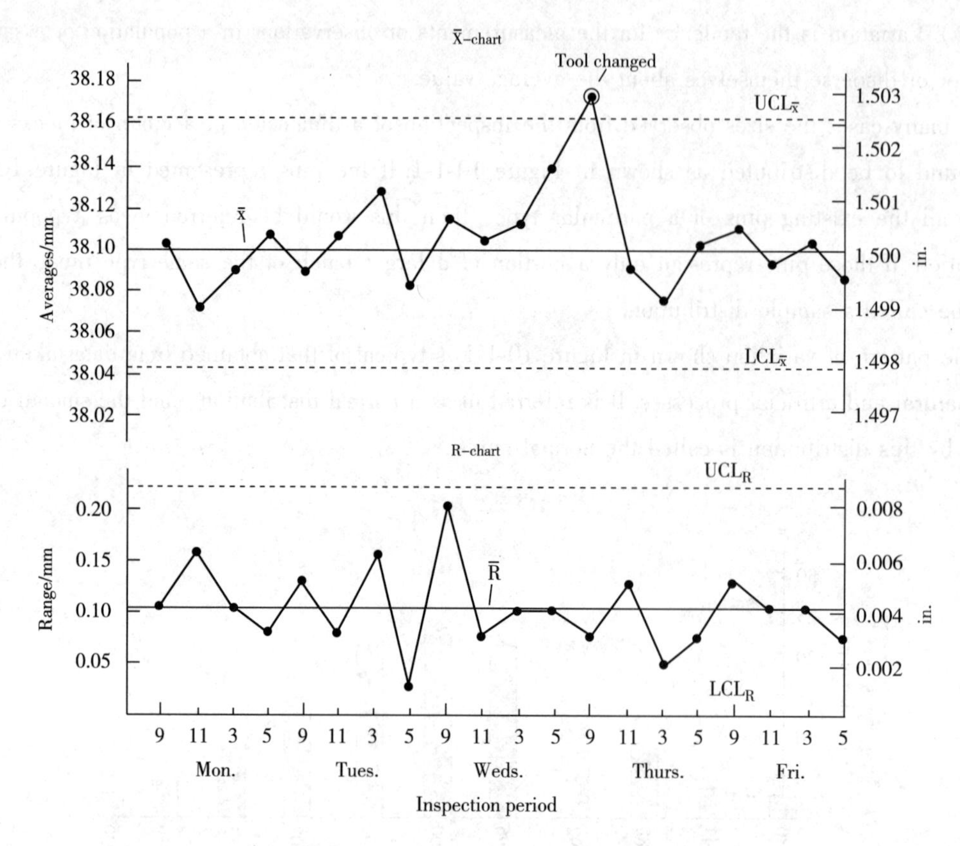

Figure 10-1-2 Quality control chart, showing averages and ranges of samples

Unit 2 Surface Quality

Text

The quality of surface finish is commonly specified along with linear and geometric dimensions. This is becoming more common as product demands increase because surface quality often determines how well a part performs. Heat-exchanger tubes transfer heat better when their surfaces are slightly rough rather than highly finished. Brake drums and *clutch* plates work best with some degree of surface *roughness*. On the other hand, bearing surfaces for high-speed engines wear in excessively and fail sooner if not highly finished but still need certain surface textures to hold lubricants①. Thus the need is to control all surface features, not just roughness alone.

Surface characteristics. Machined surface quality means the surface status of machined machinery parts. It includes mainly the following contents:

(1) Geometrical shape of surface layer.

(2) Physical and mechanical properties of surface layer, such as work-hardening of surface layer, metallurgical structure change of surface layer and *residual* stress of surface layer.

The American National Standards Institute has provided a set of standard terms and symbols to define such basic surface characteristics as *profile*, roughness, *waviness*, *flaws* and lay. Profile is defined as the contour of any section through a surface. Roughness refers to relatively finely spaced surface *irregularities* such as might be produced by the action of a cutting tool or grinding wheel during a machining operation②. Waviness consists of those surface irregularities which are of greater spacing than roughness. Waviness may be caused by vibrations, machine or work *deflections*, *warping*, etc. Flaws are surface irregularities or *imperfections* which occur at infrequent intervals and at random locations. Such imperfections as *scratches*, *ridges*, holes, *cracks*, *pits*, checks, etc., are included in this category. Lay is defined as the direction of the *predominant* surface pattern. These characteristics are illustrated in Figure 10-2-1.

Surface quality specifications. Standard symbols to specify surface quality are included in Figure 10-2-1. Roughness is most commonly specified and is expressed in units of micrometers (μm), nanometers (nm) or microinches (μin.). According to the American National Standards ANSI B46. 1—1978, the standard measure of surface roughness adopted by the United States and approximately 25 other countries around the world is the *arithmetic* average roughness, R_a (formerly AA or CLA). R_a represents the arithmetic average deviation of the ordinates of profile height increments of the surface from the centerline of that surface.

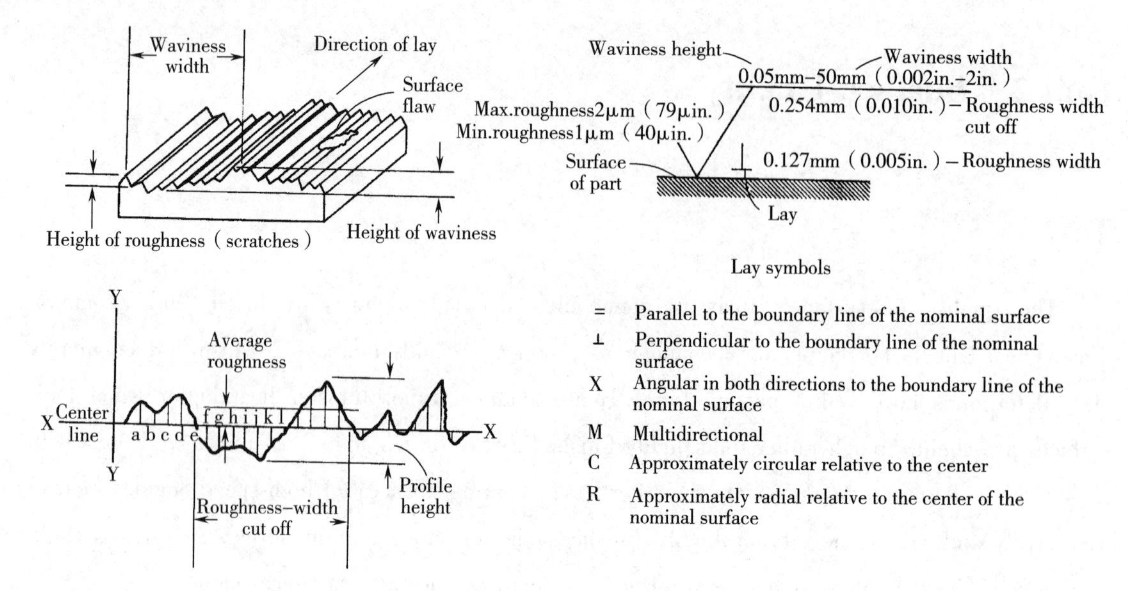

Figure 10-2-1 Upper left, typical surface highly magnified; lower left, profile of surface roughness; right, specifications of surface quality

Questions

1. Describe the standard terms of the basic surface characteristics.
2. Describe the standard symbols of the basic surface characteristics.
3. What is meant by the term "roughness"?
4. Why is surface finish important to the functional performance of mating parts?

New Words and Expressions

1. clutch [klʌtʃ] v. 抓牢 n. 把握，离合器
2. roughness [rʌfnis] n. 粗糙度
3. residual [riˈzidjuəl] adj. 残余的，有后效的 n. 剩余，残余产物
4. profile [ˈprəufail] n. 剖面，断面图，轮廓 vt. 画轮廓
5. waviness [ˈweivinis] n. 波动性，波度，波纹形
6. flaw [flɔː] n. 裂缝，裂纹，缺陷 v. 有缺陷
7. irregularity [iˌregjuˈlæriti] n. 不规则，无规律
8. deflections [diˈflekʃən] n. 偏转，偏移，偏差
9. warp [wɔːp] vt. 使翘起，歪曲 vi. 变歪 n. 歪曲
10. imperfection [impəˈfekʃən] n. 不完全，不完整，缺点
11. scratch [skrætʃ] vt. 擦，刮，涂写 n. 刮痕，擦伤
12. ridge [ridʒ] n. 波峰，隆起线 v. 使起趋
13. crack [kræk] vt. 弄裂 vi. 断裂 n. 裂缝
14. pit [pit] n. 穴，坑 vt. 使成凹

15. predominant [priˈdɔminənt] adj. 主要的，流行的，占优势的
16. arithmetic [əˈriθmətik] n. 算术，计算
17. along with 与……一道，随着，与……同时
18. in the rough 未加工，未完成，粗略

Notes

[1] On the other hand, bearing surfaces for high-speed engines wear in excessively and fail sooner if not highly finished but still need certain surface textures to hold lubricants.

另一方面，如果没有较高的表面质量，高速发动机轴承表面会过度磨损并较快失效，但是其表面还需要一定的表面组织以维持润滑。

[2] Roughness refers to relatively finely spaced surface irregularities such as might be produced by the action of a cutting tool or grinding wheel during a machining operation.

表面粗糙度指的是相对细小的不规则空间导致的表面不平整程度，像这样的不规则表面可能是在加工操作期间由于刀具或砂轮的作用而产生的。

句中"... such as..."，such 为代词，作"这样的人、事、物"解，as 从句是定语从句，说明 such。

Glossary of Terms

1. surface quality 表面质量
2. surface finish (smoothness) 表面光洁度
3. surface characteristic 表面特性
4. surface defect 表面缺陷
5. surface structure 表面构造
6. surface texture 表面组织
7. surface roughness (asperity) 表面粗糙度
8. surface quality measurement 表面质量测定
9. surface hardening (face hardening) 表面硬化
10. residual stress 残余应力
11. surface on the workpiece 工件表面
12. work surface 待加工表面
13. machined surface 已加工表面
14. transient surface 过渡表面
15. working reference plane 工作基面
16. geometrical shape of surface 表面几何形状
17. physical and mechanical properties of surface layer 表面层的物理和化学性能
18. metallurgical structure change 冶金组织变化
19. measurement method for vibration 振动测量方法
20. testing of the geometric accuracy 几何精度检验

21. testing of the working accuracy 工作精度检验
22. scratch hardness 擦硬度，刻划硬度
23. acceptance conditions and acceptance tests 验收条件和验收检验
24. flaw detector 探伤仪

Reading Materials

Measurement of Surface Finish

Waviness and roughness are measured separately. Waviness may be measured by sensitive dial indicators. A method of detecting gross waviness is to coat a surface with a high-gloss film, such as mineral oil, and then reflect in it a regular pattern, such as a wire grid. Waviness is revealed by irregularities or discontinuities in the reflected lines.

Many optical methods have been developed to evaluate surface roughness. Some are based on interferometry. One method of interference contrast makes different levels stand out from each other by lighting the surface with two out-of-phase rays. Another method projects a thin ribbon of light at 45° onto a surface. This appears in a microscope as a wavy line depicting the surface irregularities. For a method of replication, a plastic film is pressed against a surface to take its imprint. The film then may be plated with a thin silver deposit for microscopic examination or may be sectioned and magnified. These are laboratory methods and only economical in manufacturing where other means are not feasible, as on a surface inaccessible to a probe.

Surface Roughness, Waviness and Lay

The modern demands of the automobile, the airplane, and other modern machines that can stand heavier loads and higher speeds with less friction and wear have increased the need for accurate control of surface quality by the designer regardless of the size of the feature. Simple finish marks are not adequate to specify surface finish on such parts.

Surface finish is intimately related to the functioning of a surface, and proper specification of finish of such surfaces as bearings and seals is necessary. Surface quality specifications should be used only where needed, since the cost of producing a finished surface becomes greater as the quality of the surface called for is increased. Generally, the ideal surface finish is the roughest one that will do the job satisfactorily.

The system of surface texture symbols recommended by ANSI/ASME (Y14.36M-1996) for use on drawings, regardless of the system of measurement used, is now broadly accepted by American industry. These symbols are used to define surface texture, roughness and lay. See Table 10-2-1 for the meaning and construction of these symbols. The basic surface texture symbol in Figure 10-2-2a indicates a finished or machined surface by any method, just as does the general V symbol. Modifications to the basic surface texture symbol (Table 10-2-1b ~ d) define restrictions on material removal for the finished surface. Where surface texture values other than roughness average (R_a) are specified, the symbol must be drawn with the horizontal extension, as shown in Table 10-2-1e. Construction details for the symbols are given in Table 10-2-1f.

Applications of the surface texture symbols are given in Figure 10-2-2a. Note that the symbols read from the bottom and/or the right side of the drawing and that they are not drawn at any angle or upside down.

Measurements for roughness and waviness, unless otherwise specified, apply in the direction that gives the maximum reading, usually across the lay (Figure 10-2-2b)

Table 10-2-1 Surface texture symbols and construction (ANSI/ASME Y14. 36M—1996)

Symbol	Meaning
a)	Basic Surface Texture Symbol. Surface may be produced by any method except when the bar or circle, b) or d), is specified
b)	Material Removal By Machining is Required. The horizontal bar indicates that material removal by machining is required to produce the surface and that material must be provided for that purpose
c) 3.5	Material Removal Allowance. The number indicates the amount of stock to be removed by machining in millimeters (or inches). Tolerances may be added to the basic value shown or in a general note
d)	Material Removal Prohibited. The circle in the vee indicates that the surface must be produced by processes such as casting, forging, hot finishing, cold finishing, die casting, powder metallurgy or injection molding without subsequent removal of material
e)	Surface Texture Symbol. To be used when any surface characteristics are specified above the horizontal line or to the right of the symbol. Surface may be produced by any method except when the bar or circle, b) or d), is specified

f)

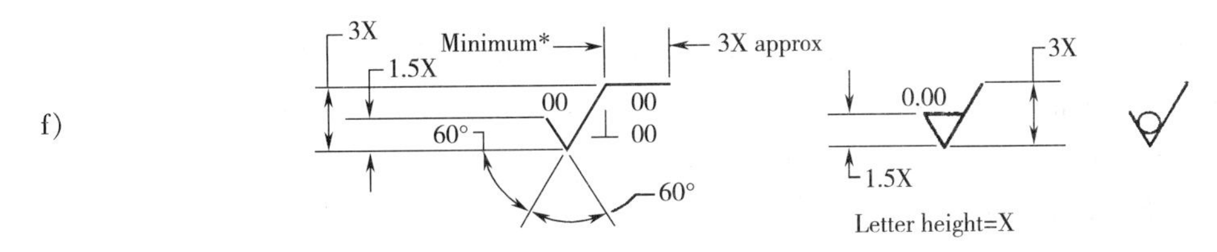

* This dimension is adjusted by + 1 for each line of values beyond the two lines shown below the horizontal line

The Effect of Various Technical Factors on Machined Surface Quality

1. Cutting tools, dies and molds wear. Thus part dimensions vary over a period of time.

2. Machinery performs differently depending on its age, condition and maintenance. Thus older machines tend to vibrate, are difficult to adjust, and do not maintain tolerances as well as new machines do.

3. Metalworking fluids perform differently as they degrade. Thus surface finish of the workpiece, tool life and forces are affected.

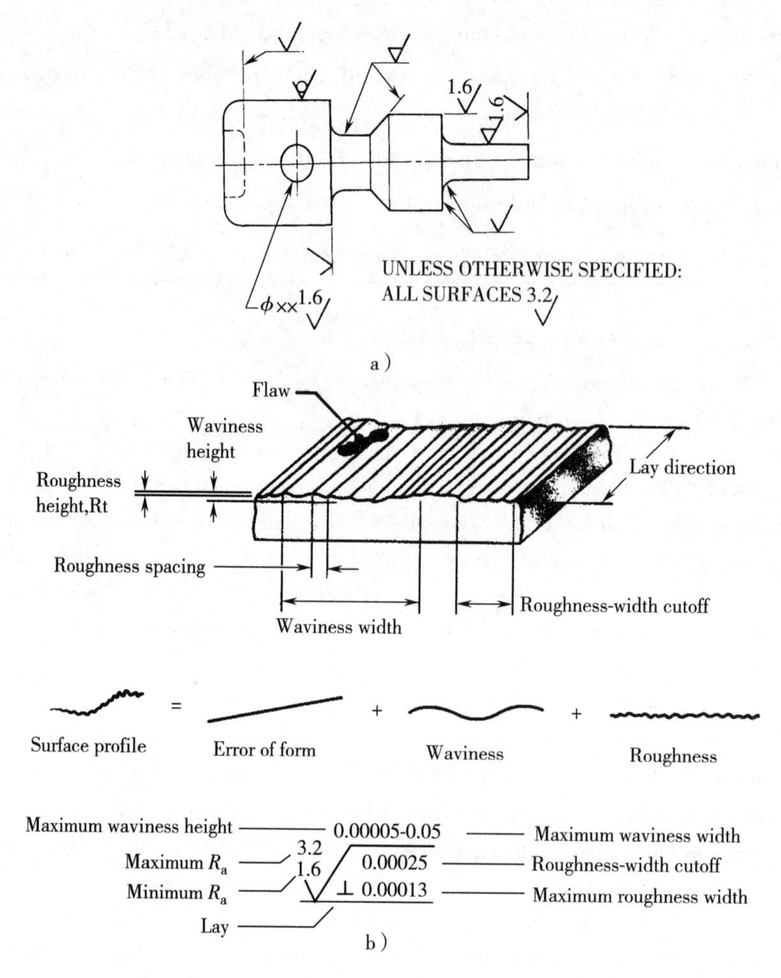

Figure 10-2-2 Application of surface texture symbols and surface characteristics (ANSI/ASME Y14. 36M-1996)

4. Environmental conditions, such as temperature, humidity and air quality in the plant may change from one hour to the next, affecting machines, workpieces and employees.

5. Different shipments of raw materials may have significantly different dimensions, properties and surface characteristics.

6. Operator skill and attention may vary during the day, from machine to machine, or among operators.

In the preceding list, those events that occur randomly, that is, without any particular trend or pattern, are called chance variations. Those that can be traced to specific cause are called assignable variations. Examples of the first type are equipment accuracy, weather effects and material properties. The second type includes factors such as worn-out equipment, out-of-specification machine adjustments, improper tooling, material defects and human errors.

Unit 3　Measuring and Gaging Instruments

Text

The dimensions of manufactured parts have to be checked as they are made and after they are finished to assure quality. Measuring is done to find the actual size of a dimension while *gaging* merely shows whether a dimension is within specified limits. Parts made in large quantities are gaged because that takes the least time to *sort* the bad from the good. Less skill is needed for gaging, as a rule, than for measuring.

Measuring instruments and gages may be classified as *precision* and non-precision. The precision of measurement of an instrument is the smallest *increment* of size that it can *reveal.* Instruments capable of measuring to 0.02mm (or 0.001 in.) or less are usually considered precision measuring instruments. The need for such devices arises from the fact that the tolerances on the dimensions of many parts manufactured today are less than 25μm or 0.001 in. and not uncommonly as small as 25nm or 0.000001 in. Measuring instruments may be direct reading or of the transfer type.

Most of the available measuring instruments may be grouped according to certain basic principles of operation. Many simple instruments use only a graduated scale as a measurement basis, while others may have two related scales and use the vernier principle of measurement. In a number of instruments the movement of a precision screw is related to two or three graduated scales to form a basis for measurement. Many other instruments utilize some sort of mechanical, electrical, or optical *linkage* between the measuring element and the graduated scale so that a small movement of the measuring element produces an enlarged indication on the scale[①]. Air pressure or metered airflow is used in a few instruments as a means of measurement. These operating principles will be more fully explained in the descriptions of a few of the instruments in which they are applied.

Basic measuring instruments and gages. People in nearly every craft, whether they be television repairmen, automotive mechanics, or telephone installers, have a set of tools applicable to their work[②]. Similarly, machinists, machine operators, toolmakers, etc., have a variety of basic measuring and gaging devices which they use in their daily work and which may be considered tools of their trade. Many of these are often referred to as standard or general-purpose tools, and for the most part they are relatively simple in construction. The results obtained from the use of these tools are considerably dependent upon the skill and *dexterity* of the person using them. For instance, the accuracy obtained in many cases depends upon the amount of pressure applied to the measuring elements. Thus the *craftsman* through training and experience *acquires* the sense of *touch* necessary to apply the tools properly. The group includes direct reading and transfer type linear measuring instruments, angular measuring devices, fixed and adjustable gaging devices, layout tools, and other miscellaneous metalworking *accessories.*

Questions

1. What is the difference between measuring and gaging?
2. What is the precision of measurement of a measuring instrument?
3. Explain the principle of operation of a vernier scale on a measuring instrument.
4. What is the standard measuring instrument?
5. What factors should be considered in selecting measurement equipment?

New Words and Expressions

1. gage [geidʒ] n. 标准尺寸，标准规格；量规；范围 vt. 测量
2. sort [sɔːt] n. 种类，类别 vt. 分类
3. precision [priˈsiʒən] n. 精确性，精密度 adj. 精确的，精密的
4. increment [ˈinkrimənt] n. 增加，增大，增值
5. reveal [riˈviːl] vt. 展现，揭示 n. 窗侧，门侧
6. linkage [ˈliŋkidʒ] n. 连接，联动，杠杆机构
7. dexterity [deksˈteriti] n. 灵巧，熟练
8. craftsman [ˈkrɑːftsmən] n. 工匠，技工
9. acquire [əˈkwaiə] vt. 获得，取得
10. touch [tʌtʃ] vt. 接触，碰到
11. accessory [ækˈsesəri] n. 附件 adj. 附属的
12. for instance 例如，举例来说
13. at the instance of 由于……的提议，应……请求
14. in case of 遇到……的时候，万一
15. in touch with 和……接触，和……一致
16. in some sort 多少，稍微
17. of a sort 同种类的，可以说是……的

Notes

[1] Many other instruments utilize some sort of mechanical, electrical, or optical linkage between the measuring element and the graduated scale so that a small movement of the measuring element produces an enlarged indication on the scale.

许多其他测量仪器在测量要素和刻度盘之间采用某种机械的、电的或光的联动装置，使得测量要素的微量移动体现在刻度盘上是一个放大的读数（指示）。

句中由 so that 引导出一个结果状语从句，可译为“以致于，使得，因此”。

[2] People in nearly every craft, whether they be television repairmen, automotive mechanics, or telephone installers, have a set of tools applicable to their work.

几乎在每一个行业的人们，不管他们是电视机维修工、汽车机械修理师还是电话安装员，都有一套适用于其工作的工具。

句中由 whether... or... 引导出让步状语从句，可译为“不管，还是”。

Glossary of Terms

1. measuring instrument 测量工具，计量仪器
2. measuring implement 量具
3. measuring gage 量规
4. measured value 测量值
5. measurement error 测量误差
6. measurement range 测量范围
7. measuring microscope 量度显微镜
8. measuring oscilloscope 测量示波器
9. linear measuring instrument 线性测量仪
10. direct reading measuring instrument 直接读数测量仪
11. indirect reading measuring instrument 间接读数测量仪
12. angular measuring device 角度测量装置
13. steel ruler 钢直尺
14. vernier caliper 游标（卡）尺
15. vernier depth gage 深度尺
16. vernier height gage 高度尺
17. height gauge 测高规
18. micrometer caliper 千分尺
19. gage block (bar) 块规，量块（规杆）
20. bevel protractor 角度尺，活动量角器
21. caliper gage 卡规
22. plug gage (thickness gage) 塞尺
23. thread gage 螺纹规
24. taper gage 锥形规
25. fixed-type gage 固定型量规
26. check gage 检验量规
27. standardized part 标准（零）件
28. standard measuring instrument 标准测量仪器
29. standard scale (gauge marks) 标准（刻度）尺
30. standard sample 标准试样（试棒，样品）
31. surface finish measuring device 表面粗糙度量具
32. coordinate-measuring machine 坐标测量仪（机）
33. flexible ruler 卷尺

Reading Materials

Linear Measuring Instruments

Basic direct reading measuring instruments. Linear measuring instruments are of two types:

direct reading and indirect reading. Most of the basic or general purpose of linear measuring instruments are typified by the steel ruler, the vernier caliper, or the micrometer caliper.

Steel rulers. The steel rulers are used effectively as line measuring devices, which mean that the ends of a dimension being measured are aligned with the graduations of the scale from which the length is read directly. Steel rulers are found in depth rulers, for measuring the depth of slots, holes, etc. They are also incorporated in slide calipers, where they are adapted to end measuring operations, which are often more accurate and easier to apply than in line measuring.

Verniers. The vernier caliper shown in Figure 10-3-1 typifies the type of instrument using the vernier principle of measurement. It is an end measuring instrument, available in various sizes, that can be used to make both outside and inside measurements. The main or beam scale on a typical metric vernier caliper is numbered in increments of 10mm, with the smallest scale division being equivalent to 1mm. The vernier scale slides along the edge of the main scale and is divided into 50 divisions, so that these 50 divisions are the same in total length as 49 divisions on the main scale. Each division on the vernier scale is then equal to (49/50) mm or 0. 98mm, which is 0. 02mm less than each division of the main scale. Aligning the zero lines of both scales would cause the first lines on each scale to be 0. 02mm apart, the second lines 0. 04mm apart, etc. A measurement on a vernier coincides with a line on the main scale. For example, the metric scale (top scale) of the illustration included with Figure 10-3-1 shows a reading of 12. 42mm. The zero index of the vernier is located just beyond the line at 12mm on the main scale, and line 21 (after 0) on the vernier coincides with a line on the main scale indicating the zero index is 0. 42mm beyond the line at 12mm. Thus 12. 00 + 0. 42 = 12. 42mm.

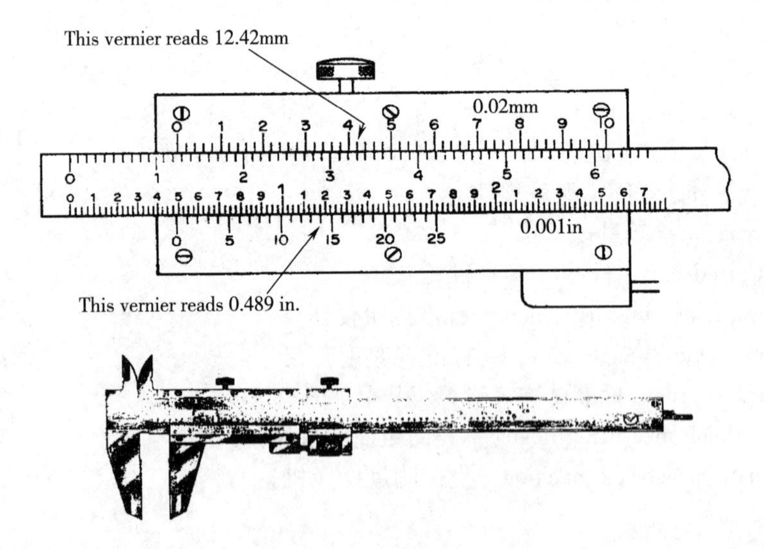

Figure 10-3-1 Vernier caliper

Micrometer. The micrometer caliper, more commonly called a micrometer, is one of the most widely used measuring devices. The micrometer illustrated in Figure 10-3-2 is representative of the type of instrument using a precision screw as a basis for measuring. It is an end measuring instrument for use in measuring outside dimensions. The measuring elements consist of a fixed anvil and a spindle that moves lengthwise as it is turned.

The thread on the spindle of a typical metric micrometer has a lead of (1/2) mm or 0.50mm, so that one complete revolution of the thimble produces a spindle movement of this amount. The graduated scale on the sleeve of the instrument has major divisions of 1.0mm and minor divisions of 0.50mm. Thus one revolution of the spindle causes the beveled edge of the thimble to move through one small division on the sleeve scale.

A reading on a micrometer is made by adding the thimble division which is aligned with the longitudinal sleeve line to the largest reading exposed on the sleeve scale. For example, in Figure 10-3-2 the thimble has exposed the number 10, representing 10.00mm, and one small division worth 0.50mm. The thimble division 16 is aligned with the longitudinal sleeve line, indicating that the thimble has moved 0.16mm beyond the last small division on sleeve. Thus the final reading is obtained by summing the three components, 10.00 + 0.50 + 0.16 = 10.66mm.

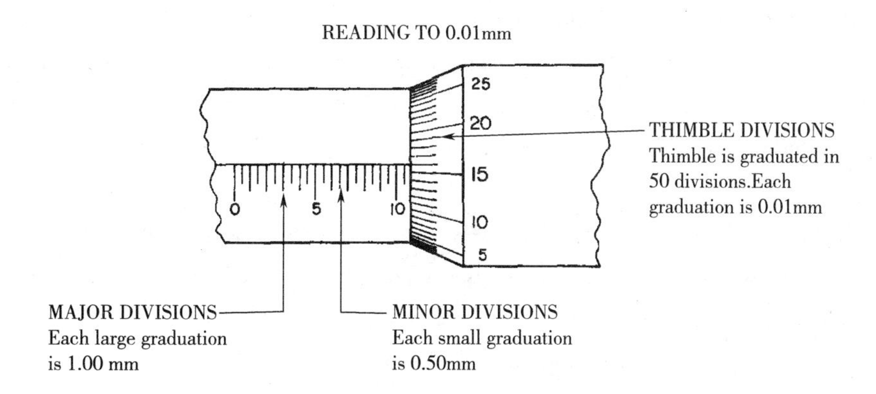

Figure 10-3-2 Micrometer reading of 10.66mm

Basic transfer-type linear measuring devices. Transfer-type linear measuring devices are typified by the spring caliper, spring divider, firm joint caliper, telescoping gage and small hole gage.

The outside caliper is used as an end measure to measure or compare outside dimensions, while the inside caliper is used for inside diameters, slot and groove widths, and other internal dimensions. They are quite versatile, but due to the construction and method of application their accuracy is somewhat limited.

Angle Measuring Instruments

The unit standard of angular measurement is the degree. The measurement and inspection of angular dimensions are somewhat more difficult than linear measurement and may require instruments of some complexity if a great deal of angular precision is required.

The bevel protractor is the most general angle measuring instrument. The two movable blades are brought into contact with the sides of the angular part, and the angle can be read on the vernier scale to five minutes of arc. A clamping device is provided to lock the blades in any desired position so that the instrument can be used for both direct measurement and layout work.

As indicated previously, the angle attachment on the combination set also can be used to measure angles in a manner similar to the bevel protractor, but usually with somewhat less accuracy.

The toolmakers' microscope is very satisfactory for making angle measurements, but its use is restricted to small parts. The accuracy obtainable is 5 minutes of arc. Similarly, angles can be measured on the contour projector.

When very accurate angle measurements are required, a sine bar may be employed if the physical conditions will permit. This consists of an accurately ground bar on which two accurately ground pins, of the same diameter, are mounted an exact distance apart. The distances used are usually either 5 in. or 10 in., and the resulting instrument is called a 5- or 10-in. sine bar. Measurements are made by using theprinciple that the sine of a given angle is the ratio of the opposite side of the right triangle to the hypotenuse.

Gages

Gage blocks. Gage blocks provide industry with linear standards of high accuracy that are necessary for everyday use in manufacturing plants. These are small, steel blocks, usually rectangular in cross section, having two very flat and parallel surfaces that are certain specified distances apart. Such gage blocks were first conceived by Carl E. Johansson in Sweden just prior to 1900.

Gage blocks usually are made of alloy steel, hardened and carefully heat-treated (seasoned) to relieve internal stresses and minimize subsequent dimensional change. Some are made entirely of carbides, such as chromium or tungsten carbide, to provide extra wear resistance. The measuring surfaces of each block are ground to approximate the required dimension and then lapped to reduce the block to the final dimension and to produce a very flat and smooth surface.

Classes of gages. In mass-manufacturing operations it is often uneconomical to attempt to obtain absolute sizes during each inspection operation. In many cases it is only necessary to determine whether one or more dimensions of a mass-produced part are within specified limits. For this purpose a variety of inspection instruments referred to as gages are employed. However, the distinction between gaging and measuring devices is not always clear as there are some instruments referred to as gages that do give definite measurements.

To promote consistency in manufacturing and inspection, gages may be classified as working, inspection, and reference or master gages. Working gages are used by the machine operator or shop inspector to check the dimensions of parts as they are being produced. Working gages usually have limits within those of the piece being inspected. Inspection gages are used by the inspection personnel to inspect purchased parts when received or manufactured parts when finished. These gages are designed and made so as not to reject any product previously accepted by a properly designed and functioning working gage. Reference or master gages are used only for checking the size or condition of other gages, and represent as exactly as possible the physical dimensions of the product.

Common gages. They include ring and snap gages for outside dimensions, plug gages for holes, and thread, form, and taper gages. Some are fixed for size, and others can be adjusted over small ranges. Some fit only one size, but others check dimension limits. Limit gages may be of the double-end type, and so that the "go" and "not go" members can be applied independently. Progressive-type gages, are quicker to use. Progressive plug gages are not applicable for blind holes. The size of a tapered piece determines how far it enters a tapered hole.

Coordinate-Measuring Machines

As schematically shown in Figure 10-3-3, a coordinate-measuring machine (CMM) basically consists of a platform on which the workpiece being measured is placed and then is moved linearly or rotated.

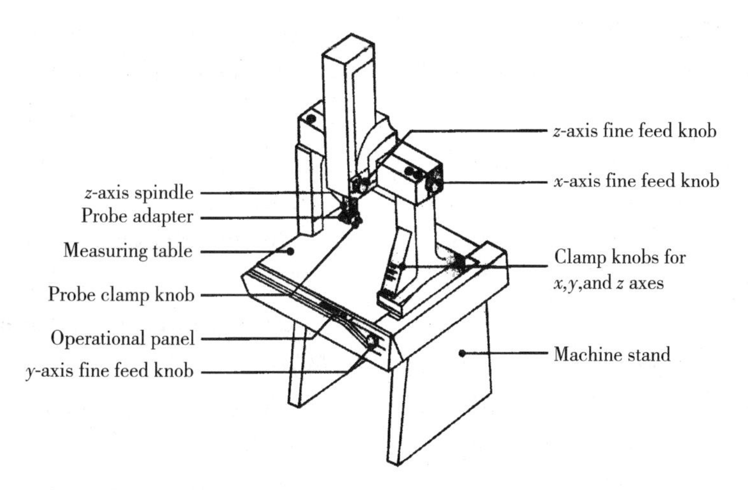

Figure 10-3-3 Schematic illustration of a coordinate-measuring machine

Coordinate-measuring machines are very versatile and capable of recording measurements of complex profiles with high resolution (0.25μm) and high speed. They are built rigidly and ruggedly to resist environmental effects in manufacturing plants, such as temperature variations and vibration. They can be placed close to machine tools for efficient inspection and rapid feedback; this way, processing parameters are corrected before the next part is made. Although large CMMs can be expensive, most machines with a touch probe and computer controlled three-dimensional movements are suitable for use in small shops.

Appendix A Standards

I. 主要国家标准和标准协会

1. GB 中华人民共和国强制性国家标准（简称国标）

2. GB/T 中华人民共和国推荐性国家标准

3. GBn 国家内部标准

4. International Organization for Standardization (ISO) 国际标准化组织

5. British Standard（BS） 英国标准

6. Japanese Standard（JS） 日本标准

7. American Standard（AS） 美国标准

8. Japanese Industrial Standard（JIS） 日本工业标准

9. European Standard（EN） 欧洲标准

10. Norme Francaise（NF） 法国标准

11. Australian Standard（AS） 澳大利亚标准

12. Indian Standard（IS） 印度标准

13. British Standards Institution（BSI） 英国标准学会

14. American Standard Association（ASA） 美国标准协会

15. American National Standards Institute（ANSI） 美国国家标准学会

16. Deutsches Institut für Normung e. V.（DIN） 德国标准化学会（德国工业标准）

17. American Iron and Steel Institute（AISI） 美国钢铁学会

18. American Society of Mechanical Engineers（ASME） 美国机械工程师学会

19. American Society of Testing and Materials（ASTM） 美国试验与材料学会

20. Chinese Welding Society 中国焊接学会

21. International Institute of Welding 国际焊接学会

22. American Welding Society（AWS） 美国焊接学会

23. British Numerical Control Society（BNC） 英国数控学会

24. American Society of Mechanical Engineers（ASME） 美国制造工程师学会

25. Forging Industry Association（FIA） 美国锻造工业协会

26. American Gear Manufacturers Association（AGMA） 美国齿轮制造商协会

27. American Foundrymen's Society 美国铸造工程师学会

28. American Electronics Association 美国电子协会

29. European Committee for Iron and Steel Standardization（ECISS） 欧洲钢铁标准化委员会

30. Association Francaise de Normalisation（AFNOR） 法国标准化协会

31. Canada Standards Association（CSA） 加拿大标准协会

32. Bureau of Indian Standard（BIS） 印度标准局

II. 金属切削机床部分国家标准

1. Metal-cutting machine tools-Terminology 金属切削机床 术语 GB/T 6477—2008

2. Machine tools—Lubrication systems 机床润滑系统 GB/T 6576—2002

3. General specifications for metal-cutting machine tools 金属切削机床 通用技术条件 GB/T 9061—2006

4. Non-traditional machines—Terminology—

Part 1: Basic terminology 特种加工机床 术语 第1部分：基本术语 GB/T 14896.1—2009

5. Non-traditional machines—Terminology—Part 2: Electro-discharge machines 特种加工机床 术语 第2部分：电火花加工机床 GB/T 14896.2—2009

6. Non-traditional machines—Terminology—Part 3: Electrochemical machines 特种加工机床 术语 第3部分：电解加工机床 GB/T 14896.3—2009

7. Non-traditional machines—Terminology—Part 4: Ultrasonic machines 特种加工机床 术语 第4部分：超声加工机床 GB/T 14896.4—2009

8. Non-traditional machines—Terminology—Part 5: Hybrid machining machines 特种加工机床 术语 第5部分：复合加工机床 GB/T 14896.5—2015

9. Non-traditional machines—Terminology—Part 6: Other non-traditional machines 特种加工机床 术语 第6部分：其他特种加工机床 GB/T 14896.6—2015

10. Metal-cutting machine tools—Method of type designation 金属切削机床 型号编制方法 GB/T 15375—2008

11. Green manufacturing—The technology specification for metal-cutting machine tool remanufacturing 绿色制造 金属切削机床再制造技术导则 GB/T 28615—2012

12. Metal-cutting machine tools—General specifications of machining parts 金属切削机床 机械加工件通用技术条件 GB/T 25376—2010

13. Metal-cutting machine tools—General specifications of assembling 金属切削机床 装配通用技术条件 GB/T 25373—2010

14. Metal-cutting machine tools—Measuring method of cleanliness 金属切削机床 清洁度的测量方法 GB/T 25374—2010

Appendix B　Translation Knowledge

Part （Ⅰ）翻译的基础知识简介

一、什么是翻译

翻译是各个不同民族之间进行跨文化的信息沟通和交流的活动。翻译本身是一种语言活动，是利用一种语言把另一种语言所表达的内容重新表达出来的过程。翻译与创作不同，翻译者不能任意地表达自己的思想，他只能准确而完整地传达原文的思想内容。

然而各民族之间由于文化背景的不同，生活习惯的差异，在语言的表达上也必然存在着千差万别。翻译即是通过不同语言的特点、规律的对比，找出相应的表达手段。

试看下面的例子：

Lathe sizes range from very little lathes with the length of the bed in several inches to very large ones turning a work many feet in length.

译：车床有大有小，小的床身只有几英寸长，大的能车削数英尺长的工件。

我们把原文与译文对比，就会看到原文是一个较长的简单句，而译文却是三个较短的单句。之所以有这样的差异，是因为英语在表达复杂概念时习惯使用长句，而汉语则很少使用长句，一般都分成几个短句进行描述，层次分明，言简意赅。

从这个例句不难看出，翻译绝不是简单地或原封不动地照搬原文的表达形式，而是有意识地选择与原文作用相等的语言手段来表达原意。从这个意义上讲，翻译是一种创造性的语言活动，是一种艺术。

二、翻译的标准

翻译的任务在于准确而完整地表达原文的思想内容，使读者对原文的思想内容有正确的理解。然而，怎样才算是准确、完整的翻译呢？要解决这个问题，就需要有一个通用的翻译标准。

谈到翻译标准，不免要提到严复在他的一部译著前言中提出的“信、达、雅”的标准。“信”就是忠实原文，“达”就是表达清楚，“雅”即是文字文雅。后来，鲁迅先生又提出了“信”和“顺”这两条标准。“信”和“顺”，即“忠实”“通顺”，今天已成为公认的两条翻译标准了。

“忠实”是指译文的思想内容必须忠实于原文，这一标准对科技英语的翻译尤为重要，因为科技作品的任务在于准确而系统地论述科学技术问题，它要求高度的准确性，特别是对术语、定义、定理和结论的翻译更应准确。

然而，忠实原文并不是形式上保持原文，而是内容上保持原文。翻译时切忌逐字死译，这样会有损原意。请看例句：

A roughing cut is usually to be followed by a finishing cut.

译：粗切削通常要被精切削跟随。

上述译文的毛病在于虽然词类或句子成分都与原文一致，保持了原文的形式，但却失去了原文的内涵。若译成“粗切之后，通常还要精切。”才算忠实原文。

翻译的第二条标准是“通顺”。“通顺”是指译文的语言形式必须符合汉语规范，翻译时要按照汉语的语法规律和习惯来遣词造句，做到通顺易懂。但要指出，“通顺”是在“忠实”的前提下，任何“顺而不信”的翻译都是不行的，属于乱译。如下面一个例子：

The foundation of the machine should not be constructed at a place of conspicuous temperature change, due to direct sunshine, excessive heat or vibration, or at a place contaminated with soil or dust particles.

译：机床的地基不应建在由于日晒、过热或受振动而引起的温度明显变化的地方，或受尘土污染的地方。

上述译文，读起来是通顺的，但仔细一推敲，就叫人不解了，为什么受振动会引起温度明显变化呢？问题就出在译文的不“忠实”上。译者把原来与 temperature change（温度变化）相并列的 excessive heat or vibration（过热或受振）误认为是与 sunshine（日晒）相并列的成分。正确的译文应是：“机床的地基不应建在受阳光直晒而引起温度明显变化的地方、过热或受振动的地方或受尘土污染的地方。”错译的原因在于译者对原文不求甚解，只满足于译文的通顺。为了防止这种现象的产生，翻译时一定要在透彻理解原文的基础之上，再力求译文的通顺。

从上面的例句可以看出，“忠实”与“通顺”这两条标准是辩证统一的。“忠实”是“通顺”的基础，译文不“忠实”，再“通顺”也毫无意义。反之，译文不“通顺”又必然影响到译文的“忠实”，因为只有译文通顺，才能更确切地表达原意。

三、翻译的过程

翻译的问题，主要在于怎样正确地理解原文和怎样确切地用译文语言表达其含义。因此，翻译的过程可分为理解和表达两个阶段。

第一阶段为理解阶段。理解阶段是通过阅读英语来掌握原作的思想内容，也是翻译的一个重要阶段。正确理解原作是翻译的基础。这就要求我们在翻译时，首先必须认真阅读原文，了解大致内容和专业范围，有时还需要查阅有关的资料，待到领会原作的精神时，才可下笔开译，这样就不致出现大错。

理解原文时，必须根据英语的语法规律和习惯去理解，把原作内容彻底弄清楚。对原作的理解应包括词汇、语法和专业内容三个方面。只有在这三方面都理解透彻，才能做出正确的表达。请看例句：

A stress is therefore set up between the two surfaces which may cause the glass to break.

错误译文：因而在引起玻璃破裂的两个表面之间产生了一个应力。

正确译文：因而在这两个表面之间产生了使玻璃破裂的应力。

前者产生错误的原因在于对原句的错误理解，把 which 引导的定语从句看成是修饰 surfaces 的，实际上 which 引导的从句是修饰 a stress 的，只是为了避免头重脚轻，保持句子平衡，才把定语从句放到了后面。另外，从专业角度看，使玻璃破裂的不可能是表面，而只能是应力。由此可以看出正确理解的重要性。

翻译过程的第二个阶段为表达阶段。表达这一阶段的任务是从汉语中选择恰当的表达手段，把已经理解了的原文内容重述出来。如果说在理解阶段必须“钻进去”，把原文内容吃透，那么在表达阶段就必须“跳出来”，不受英文原文表达形式的束缚，而根据汉语的语法规律和习惯来表达。

在表达阶段最重要的是表达手段的选择，也就是“跳出来”的问题。在正确理解原文的基础上，同一个句子可能有几种不同的译法，但译文质量并不相同。

请比较下面的译例：

Action is equal to reaction, but it acts in a contrary direction.

译文一：作用力与反作用力相等，但它向相反的方向起作用。

译文二：作用力与反作用力相等，但作用的方向相反。

译文三：作用力与反作用力大小相等，方向相反。

应该说三种译文在表达原意这一层面上都是正确的，但可以看出，在表达的简练和通顺上，后者依次要比前者好。

翻译过程虽可分为理解阶段和表达阶段，但两者不是截然分开的，而是互相联系的。

请看例句：

In fact, it may be said that anything that is not an animal or a vegetable is a mineral.

本句并不难理解，这是一个主从复合句，主句带有一个主语从句，而主语从句又带有一个定语从句。按语法分析，本句可译为：“事实上，可以说不是动物或植物的任何东西便是矿物。”但译文不够通顺，进一步分析，便可发现，“不是动物或植物”是“任何东西便是矿物”的条件。因此，可将此句译为：“事实上，可以说任何东西只要不是动物或植物，便是矿物。”这一改动使译文通顺多了，而且更符合原文意思。由此可见，透彻理解是准确表达的前提，而通过准确表达又能达到更透彻的理解。

Part（Ⅱ）翻译的基本技巧

在忠实于原文的前提下，如何摆脱原文形式的束缚，使译文符合汉语语言的规范是翻译过程中的一个重要问题。要解决这个问题，就需要掌握一定的翻译技巧。所谓技巧，就是在翻译实践中总结出来的一些一般性的规律，掌握这些规律会对翻译过程起到很大的帮助作用。下面对科技英语中常用的几种翻译技巧做简单介绍。

一、词义引申

翻译时常常遇到这样的问题，即很难从词典中找出恰当的词来表达原文中的词义。在这种情况下要使译文意思明确，用词符合汉语习惯，就需要对词义进行引申。词义的引申往往可从以下几方面来考虑。

（一）词义转译

翻译时遇到一些无法直译的词或词组，应根据上下文和逻辑关系引申翻译。

Zinc is easy to obtain from its ore.

锌容易从锌矿中提炼。（不译为“获得”）

Different systems of gases produce flames of different colours.

不同气体的组合产生不同颜色的火焰。（不译为“系统”）

（二）词义具体化

翻译时根据汉语的表达习惯，把原文中某些词义笼统的词引申为词义较具体的词。

1. 以名词 thing（东西，事情）为例

Other things being equal, iron heats up faster than aluminium.

其他条件相同时，铁比铝热得快。

There are many things that should be considered in determining cutting speed.

在确定切削速度时，有许多因素应当考虑。

2. 以动词 cut（切割）为例

The purpose of a driller is to cut holes.

钻床的功能是钻孔。

Hard metals can be easily cut with grinding wheels.

硬质金属很容易用砂轮磨削。

3. 以动词 go（走）为例

The moon turns around once itself while going around the earth.

月亮绕地球旋转一周的同时，也自转一周。

When we speak, sound waves begin to travel and go in all directions.

我们说话时，声波就开始传播，并向四面八方扩散。

（三）词义抽象化

翻译时，根据中英文的不同表达习惯，把原文中词义较具体的词引申为较抽象的词，较形象的词引申为一般的词。

Steel and cast iron also differ in carbon.

钢和铸铁的碳含量也不相同。（把“碳”抽象成“碳含量”）

The shortest distance between raw material and a finished part is casting.

铸造是把原材料加工成产品的最简单的方法。（不译为“最短的距离”）

（四）词的搭配

翻译时，要注意动词与名词、形容词与名词的搭配关系，遇到不合乎汉语搭配习惯的情况时，可以把动词或形容词的词义引申，以适应名词。

1. 引申动词适应名词

The oil also provides some cooling effect.

润滑油也起到一定的冷却作用。（不译为“提供作用”）

To be able to read machine drawings, there are a few common terms and symbols with you should be familiar.

为了能够看懂机械图，有一些常用的术语和符号是你应该熟悉的。（不译为“读机械图”）

2. 引申形容词以适应名词

With the development of electrical engineering, power can be transmitted over a long distance.

随着电气工程的发展，电力能实现远距离输送。（不译为“长距离”）

Many years ago in electrical engineering we used to differentiate between power engineering (heavy current engineering) and light current engineering.

许多年前，我们在电气工程中常把电力工程（强电工程）和弱电工程区分开来。（不译为“重电工程”和“轻电工程”）

二、词量的增减

由于英汉两种语言的表达方式不同，词汇的一一对应并不意味着翻译的准确。在原文与译文之间有一些词量的增减，反倒能使表达更清晰、准确。

但词量的增减绝不是无中生有，也不是任意删减，而是要根据需要，视具体情况，在译文中减去多余的词，或增加必要的词。下面就翻译过程中几种词量的增减情形进行说明。

（一）词量的增加

1. 在抽象名词后加名词

Other flight going past Venus, and also Mars, are planned.

经过金星的，还有经过火星的其他飞行方案也计划出来了。

（比较：经过金星的，还有经过火星的其他飞行也计划出来了。）

Were there no electric pressure in conductor, the electron flow would not take place in it.

导体内如果没有电压，便不会产生电子流动现象。（增加“现象”）

2. 在形容词前面加名词

In most cases small lathes are as complete as large lathes, only smaller and lighter.

大部分情况下，小车床的结构与大车床的结构一样完备，只是体积小些，重量轻些。（增加“体积”“重量”）

According to Newton's Third Law of Motion, action and reaction are equal and opposite.

根据牛顿运动第三定律，作用力和反作用力是大小相等、方向相反的。（增加“大小”“方向”）

3. 增加连贯语气的词

Manganese is a hard, brittle, gray-white metal.

锰是一种灰白色的、又硬又脆的金属。

（比较：锰是一种硬的、脆的、灰白的金属。）

In general, all the metals are good conductors, with silver the best and copper the second.

一般来说，金属都是良导体，其中以银为最好，铜次之。（增加“其中”）

4. 增加概括词

The chief effects of electric currents are the magnetic, heating and chemical effects.

电流的主要效应有磁效应、热效应和化学效应三种。（增加“三种”）

The frequency, wave length and speed of sound are closely related.

声音的频率、波长和速度三者是密切相关的。（增加“三者”）

（二）词量的减少

1. 省略冠词

Any substance is made up of atoms whether it is a solid, a liquid, or a gas.

任何物质，不管它是固体、液体还是气体，都是由原子组成的。（省略三个不定冠词）

Practically every river has an upper, a middle, and a lower part.

几乎每条河都有上游、中游和下游。（省略三个不定冠词）

2. 省略代词

By the word “alloy” we mean “mixture of metals”.

用“合金”这个词来表示“金属的混合物”。（省略人称代词 we）

Hardened steel is difficult to machine, but having been annealed, it can be easily machined.

淬火钢很难加工，但经过退火后便很容易加工。（省略人称代词 it）

3. 省略介词

Most substances expand on heating and contract on cooling.

大多数物质都热胀冷缩。(省略两个介词 on)

The critical temperature is different for different kinds of steel.

不同的钢，其临界温度各不相同。(省略介词 for)

4. 省略连词

Metal as well as non-metal expands when heated.

金属以及非金属受热都膨胀。(省略连词 when)

If there were no heat-treatment, metals could not be made so hard.

没有热处理，金属就不会变得如此硬。(省略连词 if)

5. 省略动词

The medium carbon steel has a high melting point.

中碳钢熔点高。(省略动词 has)

Stainless steels possess good hardness and high strength.

不锈钢硬度大、强度高。(省略动词 possess)

In conduction and convection, energy transfer through a material medium is involved.

在传导和对流时，能量通过某种介质进行传递。(省略动词 is involved)

Therefore, he must have a knowledge of what these properties are and mean.

因此，他必须了解这些性能是什么，意味着什么。(省略动词 have)

6. 省略名词

The size of the mould should be a little larger as the casting shrinks as it cools.

因为铸件冷却时收缩，所以铸模应该大一些。(不译为“铸模尺寸”)

Going through the process of heat treatment, metals become much stronger and more durable.

金属经过热处理后，强度更高，更加耐用。(不译为“热处理过程”)

三、词类转换

不同语系的语言，在词汇方面会有很大的不同。同一意思在不同的语言中可用不同的词类来表达。如：

It is possible to convert energy from one form into another.

可以把能量从一种形式变为另一种形式。

现将翻译中词类转换的一些规律分述如下：

(一) 译成汉语动词

1. 名词译成动词

It is man who plays the leading role in the application of electronic computers.

在使用电子计算机时，起主要作用的是人。

(比较：在电子计算机的使用方面，起主要作用的是人。)

By the use of ultrasonic waves, one can find out if there is a flaw in the metal.

人们使用超声波能够发现金属是否有裂缝。

2. 形容词译成动词

Generally speaking, neither gold nor stone are soluble in water.

一般来说，金子和石头都不能溶于水。

Steel is widely used in engineering, for its properties are most suitable for construction purposes.

钢广泛地应用于工程中，因为钢的性能非常适合于建筑使用。

3. 介词译成动词

Heat is produced when work is done against friction.

当克服摩擦而做功时，会产生热。

The letter E is commonly used for electromotive force.

通常用E这个字母表示电动势。

（二）译成汉语名词

1. 动词译成名词

The earth on which we live is shaped like a ball.

我们居住的地球，形状像一个大球。

Wrought irons behave differently from iron which contains a lot of carbon.

熟铁的性能不同于含碳量多的生铁。

2. 形容词译成名词

The more carbon the steel contains, the harder and stronger it is.

钢的碳含量越高，强度和硬度就越大。

The cutting tool must be strong, tough, hard, and wear resistant.

刀具必须有足够的强度、韧性、硬度，而且要耐磨。

（三）译成汉语形容词

1. 副词译成形容词

The electronic computer is chiefly characterized by its accurate and rapid computations.

电子计算机的主要特点是计算准确而快速。（动词 characterize 译成名词，副词 chiefly 译成形容词）

Gases conduct best at low pressures.

气体在低压下导电性最佳。

2. 名词译成形容词

In certain cases friction is an absolute necessity.

在一定的场合下，摩擦是绝对必要的。

That physical experiment was a success.

那项物理实验是成功的。

四、成分转换

由于表达方式不同，翻译时往往需要改变原文的语法结构。所采用的方法除了词类转换之外，还有句子成分的转换。如：

There are four seasons in a year.

一年有四季。

下面就几种情形加以介绍：

（一）介词宾语译成主语

The electric arc may grow to an inch in length.

电弧长度可以增加到 1 英寸。

The Fahrenheit scale is quite inconvenient, still it is used in England and the U. S. A.

华氏温标很不方便，但英美仍在使用。

（二）动词宾语译成主语

Levers have little friction to overcome.

杠杆要克服的摩擦力很小。

Conductors have very small resistances, and the smaller the resistance, the better the conductor.

导体的电阻很小，电阻越小，导体导电性越好。

（三）主语译成定语

Medium carbon steel is much stronger than low carbon steel.

中碳钢的强度比低碳钢大得多。

The earth acts like a big magnet.

地球的作用像一块大磁铁。

（四）定语译成谓语

Copper and tin have a low ability of combing with oxygen.

铜和锡的氧化能力弱。

Radar works in very much the same way as the flashlight.

雷达的工作原理和手电筒极为相似。

五、成分单译

有时，为使译文层次分明，简练明确，合乎汉语规范，有必要将原文中较长的句子成分分割翻译成单独的短句或短语。下面就几种技巧进行分述。

（一）主语单译

Lower temperatures are associated with lower growth rates.

温度降低，生长速度就慢下来。

In welding too much current will make it virtually impossible to stay the deposited metal in the right place.

焊接时，电流过大，实际上就会使熔敷金属无法停留在正确的位置上。

（二）谓语单译

It goes without saying that oxygen is the most active element in the atmosphere.

不言而喻，氧是大气中最活泼的元素。

The temperature at which the melting of a substance takes place is known to be the melting point of the substance.

众所周知，物质熔化时的温度便是该物质的熔点。

（三）定语单译

Hydrogen is the lightest element with an atomic weight of 1.008.

氢是最轻的元素，其原子量为 1.008。

Pig iron is an alloy of iron and carbon with carbon content more than 2 percent.

生铁是一种铁碳合金，其碳的质量分数在 2% 以上。

（四）状语单译

Titanium is strong enough to withstand heavy loads at high temperature.

钛有足够的强度，在高温下能承受重荷。

Heat is required to change ice to water.

冰变为水，就需要热。

六、否定译法

否定译法是指翻译时突破原文形式，采用变换语气的方法处理词句，把肯定的译成否定的，或把否定的译成肯定的。下面介绍几种这样的译法。

（一）肯定译否定

The influence of temperature on the conductivity of metals is slight.

温度对金属的导电性影响不大。

Metals, generally, offer little resistance and are good conductors.

通常，金属几乎没有电阻，因而是良导体。

（二）否定译肯定

Metals do not melt until heated to a definite temperature.

金属要加热到一定的温度才会熔化。

There is no material but will deform more or less under the action of forces.

所有材料在力的作用下多少会有些变形。

Part（Ⅲ）几种具体问题的具体译法

在这一部分里，将要介绍几种科技英语中具有特殊性的具体问题的具体译法。

一、被动语态的译法

科技英语中，被动语态的使用非常广泛。这是因为把科技问题放在句子的主语位置上，更能引起人们的注意。此外，被动语态相较于主动语态主观色彩更少，这是科技作品所需要的。因此，凡是在不需要或不可能指出行为主体的场合，或者在需要突出行为客体的场合都使用被动语态。汉语中虽然也有被动句的使用，但使用范围要狭窄得多。所以，在翻译被动句时，要做一些改变，以适应汉语的习惯。下面介绍几种常见译法。

（一）译成汉语被动句

1. 在谓语前加“被”字

Energy from different sources has been used to do useful work.

各种能源均被用来做有用的功。

Once the impurities have been removed, the actual reduction of the metal is an easy step.

杂质一旦被去除掉，真正的金属还原就开始了。

2. 在谓语前省略“被”字

Plastics have been applied to mechanical engineering.

塑料已应用于机械工程。

Metals may be cast into various shapes.

金属可铸成各种形状。

3. 在行为主体前加“被”“由”“受”“为……所”等字

The magnetic field is produced by an electric current.

磁场由电流产生。

Air is attracted by the earth as every other substance.

空气像任何其他物质一样被地球吸引着。

4. 译成“是……的”结构

Iron is extracted from the ore by means of the blast furnace.

铁是用高炉从铁矿石中提炼出来的。

Many casting defects are caused by expansion properties of sand.

许多铸造缺陷是由砂子的膨胀特性所造成的。

（二）译成汉语主动句

1. 译成有主语的主动句

Casting quality is influenced by a number of factors such as the nature of metal or alloy cast, properties of mould materials used and the casting process.

许多因素能影响铸件的质量，例如：铸造用的金属或合金的性质、所使用的模具材料的性质和铸造方法等。

It was also said earlier that NC is the operation of machine tools by numbers.

大家早就听说，数控就是采用数字技术操纵机床。

2. 译成无主句

Efficiency is usually expressed as a percentage.

通常用百分数来表示效率。

The temperature is lowered so that water may be turned into ice.

把温度降低，使水变成冰。

Care is to be taken to remove all the impurities.

要注意除去所有的杂质。

二、名称数的译法

英语中的名词具有单复数的概念，尤其是可数名词的复数，绝大部分具有词形上的变化。而汉语名词却没有单复数的明确概念。这就要求我们在翻译时，须将英语中单复数的概念表达出来。下面介绍几种译法。

（一）译成笼统的多数

If the reaction took hours, and not seconds, the fuel costs would be prohibitive.

假如这一反应需要数小时而不是几秒钟，那么燃料成本就太高了。

（二）译成具体的多数

Before turning a workpiece on the lathe, the lathe centers are to be aligned.

在车床上切削工件之前，车床两顶尖必须对准。

（三）译成叠词形式

The sun and stars are luminous bodies.

太阳和星星是发光体。

（四）译成不同种类

The properties of steels depend on the quantity of carbon they contain.

各种钢的性能取决于它们的含碳量。

三、there be 句型的译法

科技英语中的 there be 句型较日常英语中的要复杂一些，因为它往往不带状语，而带一些不同结构的定语，这使翻译相对困难一些，因此通常不译为“有……”。下面对它的几种译法加以介绍。

（一）译成有主语的“有”

For conduction and convection there must be molecules.

传导和对流必须有分子。

There are about seventy metallic elements.

金属元素大约有 70 种。

（二）译成无主语的“有”

There is some carbon dioxide in the air.

空气中有些二氧化碳。（先译地点状语，再把原主语译成“有”的宾语）

There are two kinds of computers — digital and analog.

有两类计算机，即数字式和模拟式。

There are some metals which possess the power to conduct electricity and the ability to be magnetized.

有些金属具有导电能力和被磁化的能力。

四、否定结构的译法

英语和汉语这两种语言在表达否定时，其词汇、语法，甚至语言逻辑都有很大的差别，需要我们在翻译时多加注意，否则会弄出大错。如：

All metals are not good conductors.

并非所有的金属都是良导体。（不可译为“所有的金属都不是良导体。”）

下面就翻译英语否定结构时应注意的几个问题举例说明。

（一）否定成分的转译

对于一些形式上一般否定（谓语否定），而意义上特指否定（其他成分否定）；或反过来，形式上特指否定，而意义上一般否定的句子，翻译时要按汉语习惯进行转译。

No body can be set in motion without having a force act upon it.

如果没有力作用于物体上，就不能使物体运动。

We do not consider melting or boiling to be chemical changes.

我们认为熔化或沸腾不是化学变化。

（二）部分否定的译法

由 all，every，both，always 等表示全体意义的词与否定词 not 构成的句子，所表达的意义不是全部否定，而是部分否定。

Both of the substances do not dissolve in water.

不是两种物质都溶于水。

All metals do not conduct electricity equally well.

不是所有的金属都有同样好的导电性能。

（三）否定语气的改变

英语的否定句在翻译时并不都以否定形式表达，因为有些句子虽带有否定词，却可以表达肯定的意思。

Energy is nothing but the capacity to do work.

能就是做功的能力。

An explosion is nothing more than a tremendously rapid burning.

爆炸不过是非常急促的燃烧。

（四）否定意义的表达

英语中有很多肯定句，却表达着否定的意义。实际上，尽管一些句子中不含否定词，但其中一些词组却具有否定的意义。这样的词组有：

too... to	太……不	free from	没有
too... far	太……不	fall short of	没有达到
fail to	不能	instead of	而不是
far from	完全不	in the absence of	没有……时

下面举两个例子：

Of all metals silver is the best conductor, but is too expensive to be used in industry.

所有金属中，银是最好的导体，但成本太高，不能在工业中使用。

On freezing water expands instead of contracts.

水在结冰时不是收缩，而是膨胀。

（五）双重否定的译法

1. 译成双重否定

There is no steel not containing carbon.

没有不含碳的钢。

It is impossible for heat to be converted into a certain energy without something lost.

热量转换成其他能而没有损耗是不可能的。

2. 译成肯定

There is no law that has not exceptions.

凡是规律皆有例外。

One body never exerts a force upon another without the second reacting against the first.

一个物体对另一个物体施加作用力必然会受到其反作用力。

五、强调句型的译法

在英语中，有一种强调句型“It is (was) ... that”。它是用来强调除谓语之外的主语、宾语、状语部分的。这一强调句型是把需要强调的部分放在“It is (was)”的后面，而把句子的其余部分放在“that”的后面，其特点是去掉“It is (was)”和“that”后的其余成分仍能组成一个完整的句子。翻译时常常在被强调成分之前加上“正是”“就是”等词。具体译法有以下两种。

（一）保留原文结构

It is heat that causes many chemical changes.

（强调主语）

正是热引起许多化学变化。

It is in industry that metals have found their greatest use.

（强调状语）

正是在工业中金属得到了最大限度的应用。

（二）改变原文结构

It is man that plays the leading role in the application of electronic computers.

（强调主语）

在应用电子计算机时，起主要作用的是人。

It is the losses caused by friction that we must try to overcome by various means.

（强调宾语）

我们必须尝试通过各种办法来克服的正是由摩擦引起的损失。

六、倍数的译法

在科技英语中，倍数的表达方式与汉语之间也存在差别，翻译时必须慎重处理。现将常见的倍数译法分述如下。

（一）“几倍于”的译法

1.“倍数 + as ... as”的译法

“n times + as ... as”结构可翻译成“是……的 n 倍”“为……n 倍”。

The oxygen atom is nearly 16 times as heavy as the hydrogen atom.

氧原子的重量几乎是氢原子的16倍。

The thermal conductivity of metals is as much as several hundred times that of glass.

金属的热导率是玻璃的数百倍。

2.“倍数 + 比较级 + than”的译法

“n times + 比较级 + than”的结构可翻译成“是……的 n 倍”。

Iron is almost three times heavier than aluminum.

铁的重量几乎是铝的 3 倍。

The volume of the earth is 49 times larger than that of the moon.

地球的体积是月球的49倍。

3.“倍数 + 名词（代词 that）”的译法

“n times +名词（代词 that）”的结构可翻译成“是……的 n 倍”。

The peak value of an alternating current is 1. 414 times its effective value.

交流电的峰值为其有效值的1. 414倍。

In this workshop the output of July was 3. 5 times that of January.

这个车间 7 月份的产量是 1 月份的3. 5倍。

（二）“增加几倍”的译法

英语中表示“增加几倍”时，常用“具有增加意义的动词 + 倍数”，或只用具有倍数意义的动词。翻译为“增加了（n－1）倍”、“增加（n－1）倍”或“增加到 n 倍”。

1.“表示增加意义的动词 + 倍数”

The speed exceeds the average speed by a factor of 2. 5.

该速度超过平均速度1. 5倍。

Since the voltage is increased one hundred fold, the current drops by the same proportion to one-hundredth.

由于电压增加到100倍，电流按同样比例下降到1%。

2. “表示倍数意义的动词+宾语”

表示倍数的动词有double（两倍）、treble（三倍）、quadruple（四倍）等。

As the high voltage was abruptly trebled, all the valves burnt.

由于高压突然增加了两倍，所有管子都烧坏了。

If the resistance is doubled without changing the voltage, the current becomes only half as strong.

如果电压不变，电阻增加1倍，电流就减小1/2。

（三）“减少程度”的译法

英语中还可以用“具有减少意义的词+倍数”来表示减少的数量，句中的“倍数”是指原量是现量的倍数。由于汉语不能说“减少（了）多少倍”，所以，要译成“减少（了）(n-1)/n”（指减去部分），或译成“减少到1/n”（指剩下部分）。

By using this new process the loss of metal was reduced four times.

采用这种新工艺使金属损耗量减少了3/4。

Aluminum is almost more than three times as light as copper.

铝几乎比铜轻2/3多。

七、长句的分析与翻译

长句是英语中一种常见的语言现象，但句子过长会造成理解和翻译上的困难。要知道再长的句子也是由基本结构扩展而成的，了解基本结构的扩展是翻译长句的基础。一般英语句子扩展的方式主要有以下三种。

（一）增加句子的修饰语

请看例句：

Applied science, on the other hand, is directly concerned with the application of the working laws of pure science to the practical affairs of life, and to increasing man's control over his environment, thus leading to the development of new techniques, processes and machines.

另一方面，应用科学则直接研究如何将理论科学中的定律用于生活实践，用于加强人类对周围世界的控制，从而导致新技术、新工艺和新机器的开发。

本句很长，但是个简单句。如果把句子进行简化，就会看到其基本结构并不复杂，只不过是“Applied science is directly concerned with the application of... to... and to...”。

（二）增加并列成分或并列句

请看例句：

The kinds of materials to be machined, variation in the types of machine tools chosen for an operation, the speeds at which the machining is performed, the selection of roughness and feeds, the kinds of cut (whether light or heavy), and the kinds of fluids used for cooling and lubricating are factors determining which material, of the several available, is best suited to the purpose in mind.

待加工材料的种类、某种加工所选择的机床种类、加工时的速度、粗糙度和进给量的选择、切削的方式（是轻切削还是重切削），以及冷却和润滑液的种类，这些都是确定（刀具）材料的因素。即根据这些因素来确定在几种可用材料中哪一种最适合目标要求。

本句很长，有6个并列主语，且每个主语又带有不同形式的定语。虽然句子很长，但其基本结构却很简单，即“并列主语＋ are factors determining...”。

（三）由短语或从句充当句子成分

请看例句：

Another is to enable the machine to perform its operations successively or even simultaneously on the same workpiece without any intermediate handing.

另一种方法是使机床在同一工件上连续地，甚至同时进行各种操作，而不需要任何中间操作。

这个句子的基本结构“Another is ＋表语”虽然很简单，但句中的表语是由一个多级短语充当的，即在不定式短“to enable...”中，不定式短语 to perform... 做不定式 to enable 的宾语补足语，两个介词短语 on the same workpiece 和 without any intermediate handing 又做不定式 to perform 的状语。

长句之所以难译，就是因为句中的某些短语和从句在汉语句子中不易安排的缘故。因此，做好从句的分译和疑问层次的安排是翻译的关键。举例如下：

1. The idea of a fish being able to generate electricity strong enough to light small bulbs, even to run an electric motor, is almost unbelievable.

鱼能发电，其强度足以点亮小灯泡，甚至能起动电动机，这简直是令人难以置信的。

（此句的难点在于主语 idea 的定语太长，将定语分译，把它分成3个短句，就容易翻译了。）

2. But man is a little like a detective arriving at the scene millions or billions of years after the event—and trying to reconstruct the event.

但是人类有点儿像侦探，在事情发生几百万年或几十亿年之后才到达出事现场，并且试图复述事件的经过始末。

（此句并不长，翻译的关键在于将修饰 detective 的两个分词短语 arriving... 和 trying... 不直接译成定语，而译成短句，这样译文就通顺了，也符合汉语习惯了。）

3. Small lathes with beds up to 6 ft. long and able to swing diameters up to 12 in. are commonly set on benches and are called bench lathes.

小车床通常安装在工作台上，因此称为台式车床，其床身长不超过 6 英尺，其可车削直径不超过12英寸。

（此句的要点是将主语 small lathes 的两个并列定语分别译成短句，放在句末。）

4. The reason the light from the flashlight will not bounce off a rough brick wall, as does from a smooth wooden wall, is that the light is sent off in different directions by the uneven surface of the bricks.

手电筒的光从粗糙的砖墙反射出来的情况，不像它从平滑的板壁上反射出来的那样，是由于光线被砖墙不平的表面向不同的方向反射的原因。

（此句原文的基本结构是“the reason is ＋ 表语”的简单结构，只不过 the reason 的定语从句中还包含着一个方式状语从句，使句子不太好翻译。将定语从句单译，再把主句译出，问题就迎刃而解了。）

Part（Ⅳ）科技英语疑难结构及其翻译

对原文的正确理解是翻译的基础，没有正确的理解就不会有正确的翻译。然而，影响原文正确翻译的主要因素往往是某些特殊的语法现象和某些特殊的习惯用法，统称疑难结构。对这些所谓的“疑难结构”往往不能简单地按字面意思或一般的语法规则去理解和翻译，否则容易译错。如：

The importance of proper lubrication cannot be overemphasized.

按字面意思去理解就成了“不能过分强调适当润滑的重要性”。而正确的译文却正好相反，即“适当的润滑，无论怎样强调也不会过分”。

由此可见，掌握一些疑难结构，对正确理解原文起着非常重要的作用。下面对科技英语中常见的疑难结构加以介绍。

一、As 结构

1. “as... as + 数字”

这一结构要译成“…… 达 ……”或“…… 至 ……”。

The temperature at the sun's center is as high as 10，000，000℃.

太阳中心的温度高达 10，000，000℃。

2. “as... as + 形容词（副词）”

“as A as B”这一结构要译成“既 B 又 A”或“又 B 又 A”。

Alloys are as important as useful.

合金既有用又重要。

3. “as much (many) ... as”

这一结构要译成“……（那样多）的……”或“…… 的 ……都”。

A saturated solution contains as much solid as it can dissolve.

饱和溶液含有它所能溶解的最大量的固体。

4. “the same as”

这一结构要译成“和……一样”或“与……相同”。

The tested conditions are the same as will be encountered in use.

测验条件和使用时将要遇到的条件一样。

5. “such as”

这一结构要译成“……的一种……”或“比如，例如”。

The metric system is such as logical link between its units.

米制是各单位之间具有逻辑关系的一种度量衡制。

6. “as... , so...”

这一结构要译成“正如……一样，……也……”。

As water is the most important of liquids, so air is the most important of gases.

正如水是最重要的液体一样，空气也是最重要的气体。

7. “as (so) far as... is concerned”

这一结构要译成“就……而论”。

As far as construction is concerned, the computer is similar to the human brain.

就结构而论，计算机和人脑有相似之处。

8. “形容词（分词、副词）+ as + 主语 + 动词”

这是一种倒装结构，as 在这里相当于 though，表示让步意义，它的语气比 though 更为强烈，表现更有力。一般译成“虽然（尽管）……，但是……”。

Complicated as the problem is, the electronic brain can solve it in a short period of time.

这个问题虽然复杂，但电脑能在很短的时间内将它解决。

二、It 结构

1. “It follows that”

在这一结构中，that 从句是主语从句，it 是引导词做形式主语，follow 是不及物动词表示“归结”的意思。可译为“由此可见”、“由此得出”等。

It follows that the greater conductance a substance has, the less the resistance is.

由此得出，一种物质的电导率越大，电阻越小。

2. “It is + 形容词 + for... + 不定式”

在这一结构中，it 是形式主语，for 以及其后的不定式构成不定式复合结构，for 引导的是不定式的逻辑主语。

It is necessary for us to know how to convert energy.

我们必须弄清楚能量是怎样转换的。

3. “It is no use + 动名词”

在这一结构中，it 是形式主语，实际主语是后面的动名词短语。

It is no use learning without practice.

学习理论而不实践是无用的。

4. “It is (was) + not until... that”

在这一结构中，it 不是形式主语，而是用来加强语气的。习惯上将该结构译为“直到……才……”。

It was not until 1886 that aluminum came into wide use.

直到1886年铝才得到广泛的应用。

5. “It is... since”

这也是一种强调时间状语的结构，但与上一种结构不同。这里 it 不是引导词，而是无人称代词，表示时间做主句的主语，连词 since 引出时间状语从句。一般译成“……已经……”或“自从……以来”。

It is ten years since the old scientist has been working at this problem.

这位老科学家研究这一课题已经整整 10 年了。

三、That 结构

That 的用法很多，这里主要介绍其引出从句及与其他词构成短语连词的一些用法。

1. “It goes without saying that”

这一结构通常被译为“不言而喻”“不容置疑”等。

It goes without saying that oxygen is the most active element in the atmosphere.

不言而喻，氧是大气中最活泼的元素。

2. “in that”

在这一结构中，in that 是固定搭配，其意义相当于 because 或 since 连接原因状语从句。它所说明的原因范围比较窄，着重指某一方面的原因，可译成“因为”“由于”“既然”等。

In that silver is expensive, it cannot be widely used as a conductor.

既然银的成本很高，它就不能广泛地被用作导体。

3. “providing that”

这一结构通常被译为“如果”“假定”等。

The volume of a gas is proportional to its absolute temperature providing that its pressure remains constant.

假定压力不变，气体的体积与其绝对温度成正比。

4. “seeing that”

在这一结构中，seeing that 是表示原因的连词。它所说明的原因意义是比较明显的，相当于 as。可译为“由于”“鉴于”等。

Seeing that the electronic computers are high in computing speed and reliable in operation, they have found wide application in computing and designing.

由于电子计算机运算速度与可靠性高，所以被广泛地应用于计算和设计领域。

5. “now that”

在这一结构中，now that 相当于连词 since，连接原因状语从句，可译为“既然”“由于”。

A concentration process is important now that the depletion of high grade ores is a possibility.

由于高品质的矿石有可能用完，富集工艺则十分重要。

四、Than 结构

Than 结构在英语中除了用来连接从句外，还可与其他词组成固定结构，常见的有以下几种。

1. “more... than”

这一结构除了比较级的用法外，另一种用法是对两种不同的性质加以比较，从而有所取舍。可把“more A than B”顺译成“是 A 而不是 B”，或逆译为“与其说是 B，不如说是 A”。

The division between the pure scientists and the applied scientists is more apparent than real.

理论科学家和应用科学家的区别只是表面上的，实际上并没有不同。

2. “more than + 数字”

这一结构一般被译为“……以上”“……多”。

More than 100 chemical elements are known to man; of these, about 80 are metals.

人类已知的化学元素有100多种，其中有 80 种是金属。

3. “less than + 数字”

这一结构一般被译为“…… 以下”“不到…… ”。

The carbon content in mild steel is less than 0. 3 percent.

低碳钢的含碳量在0. 3%以下。

4. “less. . . than”

在这一结构中, less 与 than 分开使用，意义与 less than 不同。它相当于“not so. . . as”结构，译成“比……小”“不及……”“不如……”等。

The electrons meet less resistance when the conductor is cold than when it is hot.

电子在导体冷却时遇到的阻力比导体处于热的状态时要小。

5. “other than”

这一结构一般被译为“除……之外”“不仅……而且还……”。

There are practical sources of heat energy other than the combustion of fossil fuels.

除了燃烧矿物燃料外，还有一些实用的热源。

6. “rather than”

这一结构一般被译为“而不是”。

It is for the travelers rather than the enthusiasts that the railways must be run.

运营铁路是为了旅客考虑而不是为了铁路爱好者。

7. “would rather. . . than”

在这一结构中，可把“would rather A than B”顺译为“宁愿 A 而不愿 B”，或逆译为“与其 B 倒不如 A”。

We would rather go on with the experiment than give it up.

我们宁愿把这项实验进行下去，而不愿放弃它。

8. “nothing more than”

这一结构一般被译为“只不过”或“仅仅是”。

The thermostat is nothing more than an electric switch that opens and closes itself at the proper temperature.

恒温器只不过是一种在适当温度能够自动开、关的电闸。

9. “no more than”

这一结构与上一结构相似，一般被译为“只不过”或“仅仅是”。

This first orbit nearest to the nucleus contains no more than 2 electrons.

最靠近原子核的轨道上仅仅含有两个电子。

10. “no sooner. . . than”

这一结构一般被译为“一……就”、“刚……就”等。

The push button had no sooner been depressed than the motor began to run.

按钮刚一压下，电动机就起动了。

五、其他结构

1. “what is + 比较级”

这一结构在句中起插入语的作用，可译为“而且”或“更……的是”，表示后面的内容在程度上更进一步。

What is more important, radar can do what the human eyes cannot do.

更重要的是，雷达能做人的眼睛所不能做的事。

2. "what if"

这一结构一般被译为"如果……怎么办"或"如果……那会怎样"。

What if the sun is not shining?

如果没有太阳照耀，那怎么办?

3. "all... not"

这一结构属于部分否定，一般被译为"不全是""并不都"，若译成"全都不"就错了。

All corrosion is not caused by oxidation.

腐蚀不全是由氧化作用引起的。

4. "cannot... too"

这一结构一般被译为"无论怎样……也不过分""越……越好"等。

We cannot estimate the value of modern science too much.

现代科学的价值，无论如何重视也不算过分。

We cannot be too careful in doing experiments.

我们做实验时，越小心越好。

5. "must not"

这一结构不可被译成"不必"，一般被译为"不准""不能""千万不"等。

We must not think that all molecules are as simple as those of hydrogen.

我们千万不能以为所有的分子都像氢分子那样简单。

6. "not... but"

这一结构一般被译为"不是……而是"。

The mirror at that time was made not of glass but of metal.

当时的镜子不是由玻璃制成的，而是由金属制成的。

7. "nothing but"

这一结构一般被译为"只""仅仅""只不过"等。

Early computers did nothing but compute: adding, subtracting, multiplying and dividing.

早期的计算机只能做加减乘除运算。

8. "what... but"

这一结构与上一结构同义，一般被译为"只""仅仅""只不过"等。

What is coal but a kind of product out of plants.

煤只是植物的一种产物。

9. "but for"

这一结构一般被译为"要是没有""要不是"等。

But for the heat of the sun, nothing could live.

要是没有太阳的热量，地球上什么也活不成。

10. "but that"

这一结构与上一结构同义，只不过引导的是从句而已。这一结构也被译为"要是没有""要不是"等。

But that I saw the machine, I could not have imagined how efficiently it works.

如果不是看见了这台机器，我无法想象它的效率会这样高。

11. “all but + 名词（代词）”

这一结构一般被译为“除……外都”。

All but the morning star have disappeared.

除了晨星外都不见了。

12. “all but + 形容词（动词）”

这一结构与上一结构不同，一般被译为“几乎”或“差点”。

The oxygen obtained from liquefied air is all but pure.

从液化空气中提取的氧几乎是纯的。

13. “with + 名词 + 分词”

这一结构属于分词独立结构，即带有自己逻辑主语的分词短语，这种结构主要用来表示伴随情况，一般被译为并列句。

The satellite is circling the earth, with its solar batteries being charged by the sun.（伴随情况）

卫星正围绕地球运转，其太阳能电池由太阳充电。

14. “with + 名词 + 逻辑表语”

这一结构的作用与上一种结构的作用相同。

The product is very pure with few impurities in it.（伴随情况）

这种产品很纯，内含杂质很少。

15. “名词所有格 + 动名词”

这一结构属于动名词的复合结构，即带有自己逻辑主语的动名词结构。

The metal's being workable is one of the reasons why it is so widely used in industry.

金属易加工，这是金属在工业上广泛使用的原因之一。

Appendix C　Tables of Weights and Measures

一、中华人民共和国法定计量单位（部分）

Table 1　Basic Units in the International Unit System
（国际单位制的基本单位）

Name of Quantity 量的名称	Name of Unit 单位名称	Symbol of Unit 单位符号
length　长度	metre　米	m
mass　质量	kilogramme　千克（公斤）	kg
time　时间	second　秒	s
electric current　电流	ampere　安［培］	A
thermodynamic temperature　热力学温度	kelvin　开［尔文］	K
amount of a substance　物质的量	mole　摩［尔］	mol

Table 2　Derived Units with Special Names in the International Unit System
（国际单位制中具有专门名称的导出单位）

Name of Quantity 量的名称	Name of Unit 单位名称	Symbol of Unit 单位符号	Other Form 其他表示
force　力；重力	newton　牛［顿］	N	$kg \cdot m/s^2$
frequency　频率	hertz　赫［兹］	Hz	s^{-1}
pressure；intensity；stress　压力；压强；应力	pascal　帕［斯卡］	Pa	N/m^2
energy；work；heat　能量；功；热	joule　焦［耳］	J	$N \cdot m$
power；radiant flux　功率；辐射通量	watt　瓦［特］	W	J/s
Celsius temperature　摄氏温度	degree Celsius　摄氏度	℃	—
capacitance　电容	farad　法［拉］	F	C/V
electrical resistance　电阻	ohm　欧［姆］	Ω	V/A
potential；voltage；electromotive 电位；电压；电动势	volt　伏［特］	V	W/A

Table 3 Non-international Units Stipulated by the State

（国家选定的非国际单位制单位）

Name of Quantity 量的名称	Name of Unit 单位名称	Symbol of Unit 单位符号	Conversion 换算关系
time 时间	minute 分 hour ［小］时 day 天（日）	min h d	1min = 60s 1h = 60min = 3600s 1d = 24h = 86400s
plane angle 平面角	second ［角］秒 minute ［角］分 degree ［角］度	(″) (′) (°)	 1′ = 60″ 1° = 60′
speed of rotation 旋转速度	rotation per minute 转每分	r/min	1r/min = （1/60） r/s
length 长度	nautical mile 海里	n mile	1n mile = 1，852m（航海）
speed，velocity 速度	knot 节	kn	1kn = 1n mile/h
mass 质量	ton 吨	t	1t = 1，000kg
cubic measure 体积	litre 升	L，(l)	$1L = 1dm^3 = 10^{-3}m^3$
energy 能	electron volt 电子伏	eV	$1eV \approx 1.6 \times 10^{-19}J$
acreage of land 土地面积	hectare 公顷	hm^2，(ha)	$1hm^2 = 10,000m^2$

二、法定计量单位与常见非法定计量单位的对照和换算表（部分）

Name of Quantity 量的名称	Legal Unit 法定计量单位		Common Non-legal Unit 常见非法定计量单位		Conversion 换算关系
	Name of Unit 单位名称	Symbol 符号	Name of Unit 单位名称	Symbol 符号	
Length 长度	kilometre 千米（公里）	km	—	KM	1km = 2 市里 = 0.6214 英里 mi
	metre 米	m	metre 公尺	M	1 米 = 1 公尺 = 3 市尺 = 3.2808 英尺 ft
	decimetre 分米	dm	decimetre 公寸	—	1dm = 0.1m = 3 市寸
	centimetre 厘米	cm	centimetre 公分	—	1cm = 1 公分 = 0.01m
	millimetre 毫米	mm	millimetre 公厘	m/m，MM	1mm = 1 公厘 = 0.1cm = 0.001m
	micron 微米	μm	—	—	1μm = 0.001mm
	—	—	mile 英里	mi	1mi = 1760yd = 5280ft = 1.609km

(continued)

Name of Quantity 量的名称	Legal Unit 法定计量单位		Common Non-legal Unit 常见非法定计量单位		Conversion 换算关系
	Name of Unit 单位名称	Symbol 符号	Name of Unit 单位名称	Symbol 符号	
Length 长度	—	—	yard 码	yd	1yd = 3ft = 0.9144m
	—	—	foot 英尺	ft	1ft = 12in = 30.48cm
	—	—	inch 英寸	in	1in = 2.54cm
Area 面积	square kilometre 平方千米（公里）	km^2	—	KM^2	$1km^2 = 100hm^2 = 0.3861mile^2$
	—	—	are 公亩	a	$1a = 100m^2 = 0.0247acre$
	square metre 平方米	m^2	平米，方	—	$1m^2 = 10.7639ft^2 = 1.196yd^2$
	square centimetre 平方厘米	cm^2	—	—	$1cm^2 = 0.0001m^2$
	—	—	square mile 平方英里	$mile^2$	$1\ mile^2 = 640acre = 2.59km^2$
	—	—	acre 英亩	—	$1\ acre = 4840\ yd^2 = 40.4686a$
	—	—	square yard 平方码	yd^2	$1\ yd^2 = 9\ ft^2 = 0.8361m^2$
	—	—	square foot 平方英尺	ft^2	$1\ ft^2 = 144\ in^2 = 0.093m^2$
	—	—	square inch 平方英寸	in^2	$1\ in^2 = 6.4516cm^2$
Cubic measure 体积	cubic metre 立方米	m^3	方，公方	—	$1\ m^3 = 35.3147ft^3 = 1.308\ yd^3$
	cubic centimetre 立方厘米	cm^3	—	—	$1\ cm^3 = 0.000001m^3$
	—	—	cubic yard 立方码	yd^3	$1\ yd^3 = 27ft^3 = 0.7646m^3$
	—	—	cubic foot 立方英尺	ft^3	$1\ ft^3 = 1728in^3 = 0.028m^3$
	—	—	cubic inch 立方英寸	in^3	$1\ in^3 = 16.3871cm^3$

(continued)

Name of Quantity 量的名称	Legal Unit 法定计量单位		Common Non-legal Unit 常见非法定计量单位		Conversion 换算关系
	Name of Unit 单位名称	Symbol 符号	Name of Unit 单位名称	Symbol 符号	
Mass 质量	ton 吨	t	tonne, metric ton 公吨	T	1t =1T =1000kg =0.9842UKton
	kilogramme 千克（公斤）	kg	—	KG, kgs	1kg =2.2046lb
	gramme 克	g	metric gramme 公分	gm, gr	1g =1gm =0.001kg =15.4324gr
	milligramme 毫克	mg			1mg =0.000001kg
	—	—	long ton 英吨（长吨）	UKton	1 UKton =2240lb =1016.047kg
	—	—	short ton 美吨（短吨）	sh ton, USton	1 USton =2000lb =907.185kg
	—	—	pound 磅	lb	1 lb =16oz =0.4536kg
	—	—	ounce 盎司	oz	1 oz =16dr =28.3495g
	—	—	dram 打兰	dr	1 dr =27.34375gr =1.7718g
	—	—	grain 格令	gr	1 gr =1/7000lb =64.80mg
Temperature 温度	Kelvin 开［尔文］	K	Kelvin temperature 开氏度	°K	1 K =1 °K =1℃ （absolute temperature 绝对度 = 开氏度）
	Celsius degree 摄氏度	℃	degree 度 Fahrenheit degree 华氏度	deg °F	1 deg =1K =1℃ 1 °F =5/9 K {n ℃ = [(n×1.8) +32] °F m °F = [(m−32) ×5/9] ℃}
Capacity 容积	litre 升	L (l)	litre 公升、立升	—	1 L =1 公升 =1 立升
	decilitre 分升	dL, dl	—	—	1 dL =0.1L
	millilitre 毫升	mL, ml	cubic centimetre 西西	c.c., cc	1 mL =1cc =0.001L

(continued)

Name of Quantity 量的名称	Legal Unit 法定计量单位		Common Non-legal Unit 常见非法定计量单位		Conversion 换算关系
	Name of Unit 单位名称	Symbol 符号	Name of Unit 单位名称	Symbol 符号	
Capacity 容积	—	—	bushel 蒲式耳（UK）		1（UK）bushel = 4（UK）pk
	—	—	peck　配克（UK）	pk	1 pk = 2 UKgal = 9.0922L
	—	—	gallon　加仑（UK）	UKgal	1 UKgal = 4UKqt = 4.54609L
	—	—	quart　夸脱（UK）	UKqt	1 UKqt = 2 UKpt = 1.1365L
	—	—	pint　品脱（UK）	UKpt	1 UKpt = 4UKgi = 5.6826dL
	—	—	gill　及耳（UK）	UKgi	1 UKgi = 1.4207dL

三、其他单位表示形式（部分）

Name of Quantity 量的名称	density　密度	speed　速率	pressure　压强
1	g. per cu. cm.（克/厘米^3）	rad. per sec.（弧度/秒）	lb per sq. ft.（磅/英尺^2）
2	kg. per cu. meter（千克/米^3）	rev. per min.（转/分）	lb per sq. in.（磅/英寸^2）
3	lb. per cu. in（磅/英寸^3）	ft. per sec.（英尺/秒）	kg per sq. cm.（千克/厘米^2）
4	lb. per cu. ft（磅/英尺^3）	mile per hr.（英里/时）	ton per sq. ft.（吨/英尺^2）

Appendix D Numerals, Days and Months

一、数字（Numerals）

1. 基数词（Cardinal numbers）

Cardinal numbers 基数词	Expressions 表达方式	Remarks 备注
1～12	1 one 2 two 3 three 4 four 5 five 6 six 7 seven 8 eight 9 nine 10 ten 11 eleven 12 twelve	1～12 的表达为固定搭配
13～19	13 thirteen 14 fourteen 15 fifteen 16 sixteen 17 seventeen 18 eighteen 19 nineteen	对个位数 3～9 进行变形，在其后加"-teen"
10 的整倍数	20 twenty 30 thirty 40 forty 50 fifty 60 sixty 70 seventy 80 eighty 90 ninety	对 2～9 进行变形，在其后加"ty"
21～99 （10 的倍数除外）	21 twenty-one 29 twenty-nine 32 thirty-two 38 thirty-eight 43 forty-three 54 fifty-four 65 sixty-five 76 seventy-six 87 eighty-seven 98 ninety-nine	由十位数和个位数字构成，中间用"-"相连
101～999	100 one hundred 101 one hundred and one 200 two hundred 210 two hundred and ten 367 three hundred (and) sixty-seven 906 nine hundred and six	由"百位数 + hundred + and + 末两(一)位数"构成
千的表示	1000 one thousand 2000 two thousand 2041 two thousand and forty-one	由"千位数 + thousand + and + ..."构成。注意"thousand"后不加"s"

注：英语中没有"万"，10，000 表达为"ten thousand"（十千），100，000 表达为"one hundred thousand"。此外，1，000，000表达为"one million"，1，000，000，000 表达为"one billion"。

2. 序数词（Ordinal numbers）

Ordinal numbers 序数词	Expressions 表达方式	Ordinal numbers 序数词	Expressions 表达方式	Ordinal numbers 序数词	Expressions 表达方式
1 st	first	3 rd	third	5 th	fifth
2 nd	second	4 th	fourth	6 th	sixth

(continued)

Ordinal numbers 序数词	Expressions 表达方式	Ordinal numbers 序数词	Expressions 表达方式	Ordinal numbers 序数词	Expressions 表达方式
7 th	seventh	15 th	fifteenth	30 th	thirtieth
8 th	eighth	16 th	sixteenth	40 th	fortieth
9 th	ninth	17 th	seventeenth	50 th	fiftieth
10 th	tenth	18 th	eighteenth	60 th	sixtieth
11 th	eleventh	19 th	nineteenth	70 th	seventieth
12 th	twelfth	20 th	twentieth	80 th	eightieth
13 th	thirteenth	21 st	twenty-first	90 th	ninetieth
14 th	fourteenth	22 nd	twenty-second	100 th	one hundredth

二、星期和月份（Days and Months）

1. 星期（Days of the week）

Days of the week 星期	Expressions 表达方式	Days of the week 星期	Expressions 表达方式
星期一	Monday (Mon)	星期五	Friday (Fri)
星期二	Tuesday (Tues)	星期六	Saturday (Sat)
星期三	Wednesday (Weds)	星期日	Sunday (Sun)
星期四	Thursday (Thurs)		

2. 月份（Months）

Months 月份	Expressions 表达方式	Months 月份	Expressions 表达方式
一月	January (Jan)	七月	July (Jul)
二月	February (Feb)	八月	August (Aug)
三月	March (Mar)	九月	September (Sept)
四月	April (Apr)	十月	October (Oct)
五月	May	十一月	November (Nov)
六月	June (Jun)	十二月	December (Dec)

三、数学表达式（Mathematical terms）

1. 逻辑运算（Logical Operations）

Logical Operations 运算	Expressions 表达方式	Symbols 符号
加	plus	+
减	minus	−
加减	plus or minus	±
乘	multiplication	×
除	division	÷
等于	equals	=

2. 小数（Decimals）

Decimals 小数	Expressions 表达方式
0. 5	（nought /zero /o）point five
0. 02	0（zero /nought）point o（zero /nought）two
1. 037	one point zero three seven
1. 1	one point one
1. 25	one point two five
63. 57	sixty-three（six three）point five seven

3. 分数（Fractions）

Fractions 分数	Expressions 表达方式
三分之一	one third
三分之二	two thirds
五分之三	three fifths
百分之一	one（or a）hundredth（=one percent）
千分之五	five thousandth
二十五分之三	three twenty-fifths
四分之一	a /one quarter（a /one forth）
二分之一	a /one half
四分之三	three-quarters（three-fourths）

(continued)

Fractions 分数	Expressions 表达方式
二又二分之一	two and a half
三又三分之一	three and a third
五又十分之三	five and three-tenths

注："基数词 + 序数词" 构成分数。

分子由基数词表示，分母由序数词表示。基数词大于 "1" 时，序数词要加 "s"。

四、数字比较关系（Numeral Relationships）

Relationships 关系	Expressions 表达方式	Relationships 关系	Expressions 表达方式
多于 例：多于 30	more than 30 over 30 above 30	少于 例：少于 30	less than 30 under 30 below 30
再加	30 more　再（加）30 three more days　再 3 天	数字 + odd 例：三十多	thirty odd（ten-odd） thirty and odd
约数	some thirty feet　三十英尺左右 about thirty feet　三十英尺左右 thirty feet or so　三十英尺左右 a hundred more or less　一百上下 more or less thirty pages　三十页左右	或多或少	a metre or more　一米或一米多 a long metre　一米还多 a metre or less　一米或不到一米 a long hour　足足一小时 a dozen　一打
范围	from ten to twenty　从十到二十 ten to twenty　从十到二十 between ten to twenty　十到二十之间	其他示例	five and five make ten　五加五等于十 one hundred and thirty　一百三十 hundreds（or thousands）of people 好几百人（好几千人） thousands upon thousands of people 成千上万/许许多多/无数人 hundreds of thousands of people　成千上万/许许多多/无数人
数词复数形式	tens　数十个 dozens　几十 Hundreds　几百个 millions　千百万		
含数词的习惯短语	a hundred（thousand）and one 无数的，许多的 first of all　首先，第一 fifty-fifty（half and half）平均，各一半 by halves　不完全，不彻底		

五、比例 (Scales)

Scales 比例	Expressions 表达方式
full size 比例值	1 : 1 = full
enlargements 放大倍数	2 : 1 = double size 3 : 1 = three times size 5 : 1 = five times size
reductions 缩小倍数	1 : 2 = half size 1 : 3 = third size 1 : 4 = quarter size

六、其他表示 (Other Expressions)

1) ninety degrees 90°。

2) seventy four degrees Centigrade 摄氏 74℃。

3) eighty degrees Fahrenheit 华氏 80℉。

4) six percent 6%。

References（参考文献）

[1] 曾志新，等．机械制造技术基础［M］．武汉：武汉理工大学出版社，2004.

[2] Serope Kalpakjian，Steven R Schmid. 制造工程与技术［M］.5 版．北京：清华大学出版社，2006.

[3] P N Rao. 制造技术（第 1 卷）：铸造、成形和焊接（英文版）［M］.2 版．北京：机械工业出版社，2010.

[4] P N Rao. 制造技术——金融切削与机床（英文版）［M］．北京：机械工业出版社，2003.

[5] 汪应洛．英汉机电工程技术词汇［M］．北京：科学出版社，2002.

[6] 唐一平．先进制造技术（英文版）［M］．北京：机械工业出版社，2002.

[7] 夏琴香．冲压成形工艺及模具设计［M］．广州：华南理工大学出版社，2004.

[8] 焦永和，韩宝玲，李苏红．工程图学［M］.8 版．北京：高等教育出版社，2005.

[9] 郑易里，等．英华大词典［M］.3 版．北京：商务印书馆，2000.

[10] 刘伟军，等．快速成型技术及应用［M］．北京：机械工业出版社，2005.

[11] 中国机械工程学会焊接分会，焊接词典［M］.3 版．北京：机械工业出版社，2008.

[12] 吴林，陈善本，等．智能化焊接技术［M］．北京：国防工业出版社，2000.

[13]《新英汉机械工程词汇》编订组．新英汉机械工程词汇［M］．北京：科学出版社，2002.